SpringerBriefs in Electrical and Computer Engineering

Control, Automation and Robotics

Series editors

Tamer Başar
Antonio Bicchi
Miroslav Krstic

More information about this series at http://www.springer.com/series/10198

Dimitri Breda · Stefano Maset
Rossana Vermiglio

Stability of Linear Delay Differential Equations

A Numerical Approach with MATLAB

 Springer

Dimitri Breda
Department of Mathematics and Computer
 Science
University of Udine
Udine, Italy

Rossana Vermiglio
Department of Mathematics and Computer
 Science
University of Udine
Udine, Italy

Stefano Maset
Department of Mathematics
 and Geosciences
University of Trieste
Trieste, Italy

ISSN 2192-6786
ISBN 978-1-4939-2106-5
DOI 10.1007/978-1-4939-2107-2

ISSN 2192-6794 (electronic)
ISBN 978-1-4939-2107-2 (eBook)

Library of Congress Control Number: 2014951724

Mathematics Subject Classification (2010): 34K06, 34K08, 34K13, 34K20, 34K21, 34K28, 34L16, 47D06, 65L03, 65L07, 65L15

Springer New York Heidelberg Dordrecht London

MATLAB® is owned by MathWorks.

© The Author(s) 2015

This work is subject to copyright. All rights are reserved by the Publisher, whether the whole or part of the material is concerned, specifically the rights of translation, reprinting, reuse of illustrations, recitation, broadcasting, reproduction on microfilms or in any other physical way, and transmission or information storage and retrieval, electronic adaptation, computer software, or by similar or dissimilar methodology now known or hereafter developed. Exempted from this legal reservation are brief excerpts in connection with reviews or scholarly analysis or material supplied specifically for the purpose of being entered and executed on a computer system, for exclusive use by the purchaser of the work. Duplication of this publication or parts thereof is permitted only under the provisions of the Copyright Law of the Publisher's location, in its current version, and permission for use must always be obtained from Springer. Permissions for use may be obtained through RightsLink at the Copyright Clearance Center. Violations are liable to prosecution under the respective Copyright Law.

The use of general descriptive names, registered names, trademarks, service marks, etc. in this publication does not imply, even in the absence of a specific statement, that such names are exempt from the relevant protective laws and regulations and therefore free for general use.

While the advice and information in this book are believed to be true and accurate at the date of publication, neither the authors nor the editors nor the publisher can accept any legal responsibility for any errors or omissions that may be made. The publisher makes no warranty, express or implied, with respect to the material contained herein.

Printed on acid-free paper

Springer is part of Springer Science+Business Media (www.springer.com)

To our families

Preface

Many problems of growing interest in science, engineering, biology, and medicine are modeled with systems of differential equations involving delay terms. In general, the presence of the delay in a model increases its reliability in describing the relevant real phenomena and predicting its behavior. Besides, the introduction of history in the evolution law of a system also augments its complexity since, opposite to Ordinary Differential Equations (ODEs), Delay Differential Equations (DDEs) represent infinite dimensional dynamical systems. Thus their time integration and the study of their stability properties require much more effort, together with efficient numerical methods.

Since the introduction of the delay terms in the differential equations may drastically change the system dynamics, inducing dangerous instability and loss of performance as well as improving stability, analyzing the asymptotic stability of either an equilibrium or a periodic solution of nonlinear DDEs is a crucial requirement. Several monographs have been written on this subject and the theory is well developed. By the Principle of Linearized Stability, the stability questions can be reduced to the analysis of linear(ized) DDEs. In the literature, a great number of analytical, geometrical, and numerical techniques have been proposed to answer such questions. Part of these techniques aim at analyzing the distribution in the complex plane of the eigenvalues of certain infinite dimensional linear operators, in particular the solution operators associated to the linear(ized) problem and their infinitesimal generator.

This monograph does not aim to be a survey, but presents the authors' recent work on the numerical methods for the stability analysis of the zero solution of linear DDEs, which consist in applying pseudospectral techniques to discretize either the solution operator or the infinitesimal generator. The eigenvalues of the resulting matrices are then used to approximate the exact spectra. The purpose of the book is to provide a complete and self-contained treatment, which includes the basic underlying mathematics and numerics, examples from applications and, above all, MATLAB programs implementing the proposed algorithms. MATLAB is a high-level language and interactive environment, which is nowadays well developed and widely used for a variety of mathematical problems arising from both theory and applications. Advanced students and researchers in applied mathematics, in dynamical systems, and in various fields of science and engineering concerned with delay systems are encouraged to experience the practical aspects. Having at disposal MATLAB codes to test the theory and to analyze the performances of the methods on given examples, they can tackle the numerical stability analysis of their own delay models by easily modifying these codes. Readers can also appreciate the possible application of the latter to the stability analysis of equilibria and periodic solutions of nonlinear DDEs as well as to trace bifurcation diagrams and stability charts for DDEs with varying parameters.

To furnish a solid foundation and a complete understanding of the performances of the algorithms, neither the theoretical nor the numerical analysis can be left aside. A motivated introduction to the theory of semigroups with a number of proofs is given, but the emphasis is on the (unifying) idea of using pseudospectral techniques for the numerical stability analysis of linear or linearized DDEs. Therefore, a detailed presentation of the discretization schemes is given. The monograph is completed with a fully developed error analysis, complemented with numerical results on test problems, and models from applications.

After reading the book, one should have reviewed (or acquired) the essential background on the theory of semigroups to understand the main features of the dynamical systems described by DDEs. This is the starting point for the construction of the numerical methods. Readers interested in the numerical analysis can find a complete and detailed error analysis, while readers interested more in models or applications can appreciate the role of the numerical analysis in the derivation of accurate and efficient approximations techniques. Finally, all of them should have learned how to use and modify the MATLAB codes to try new investigations (possibly reading only the first part of Chaps. 7 and 8). Eventually, such codes are made freely available, [48].

Udine, Italy, September 2014　　　　　　　　　　　　　　　　　Dimitri Breda
Trieste, Italy　　　　　　　　　　　　　　　　　　　　　　　　　Stefano Maset
Udine, Italy　　　　　　　　　　　　　　　　　　　　　　　Rossana Vermiglio

Contents

1 Introduction 1
    1.1   Linear Ordinary Differential Equations. 2
    1.2   Simple Linear Delay Differential Equations 4
    1.3   An Example from Population Dynamics 7
    1.4   An Example from Mechanical Engineering. 10
    1.5   Scopes and Synopsis 12

Part I Theory

2 Notation and Basics 17
    2.1   Notation. 17
    2.2   The Cauchy Problem. 19
    2.3   Stability of Solutions. 20

3 Stability of Linear Autonomous Equations. 23
    3.1   The Solution Operator Semigroup and the Infinitesimal
           Generator. 24
    3.2   Spectral Properties and the Characteristic Equation 27
    3.3   Linearization and Equilibria 31

4 Stability of Linear Periodic Equations 35
    4.1   The Evolution Operator and the Monodromy Operator. 36
    4.2   Linearization and Periodic Solutions 39

Part II Numerical Analysis

5 The Infinitesimal Generator Approach. 45
    5.1   The Pseudospectral Differentiation Method. 45
    5.2   The Piecewise Pseudospectral Differentiation Method 48
    5.3   Convergence Analysis 50

		5.3.1	A Related Collocation Problem.	51
		5.3.2	Convergence of the Eigenvalues	57
		5.3.3	Quadrature for Distributed Delays.	62
	5.4	Convergence of the Piecewise Method.		64
	5.5	Other Methods		64
6	**The Solution Operator Approach**			67
	6.1	The Pseudospectral Collocation Method.		68
		6.1.1	Discretization of X	68
		6.1.2	Discretization of X^+	70
		6.1.3	Discretization of T	71
	6.2	The Collocation Equation.		72
	6.3	Convergence Analysis		76
		6.3.1	Convergence of the Eigenvalues of $\widehat{T}_N$	79
		6.3.2	Convergence of the Eigenvalues of $\widehat{T}_{M,N}$	83
		6.3.3	Quadrature for Distributed Delays.	85
	6.4	Other Methods		86

Part III Implementation and Applications

7	**MATLAB Implementation**			89
	7.1	Introducing the Model in MATLAB		89
	7.2	The Infinitesimal Generator Approach		93
		7.2.1	A Single Discrete Delay	95
		7.2.2	A Single Distributed Delay	96
		7.2.3	The Piecewise Method.	97
	7.3	The Solution Operator Approach.		101
		7.3.1	The Meshes	103
		7.3.2	The Matrix $T_M^{(1)}$	105
		7.3.3	The Matrix $T_{M,N}^{(2)}$	108
		7.3.4	The Matrix $U_{M,N}^{(1)}$	110
		7.3.5	The Matrix $U_N^{(2)}$	115
8	**Applications**			117
	8.1	Test Cases		117
		8.1.1	Test 1: Linear Autonomous Equations with a Discrete Delay.	117
		8.1.2	Test 2: Linear Autonomous Equations with Multiple Discrete Delays	123
		8.1.3	Test 3: Linear Autonomous Equations with a Distributed Delay.	124

	8.1.4	Test 4: Linear Autonomous Systems	126
	8.1.5	Test 5: Linear Periodic Equations	127
	8.1.6	Test 6: Linearized Periodic Equations	129
8.2	Equilibria in Population Dynamics		130
8.3	Periodic Problems in Engineering		137

References 147

Series Editors' Biographies 157

Chapter 1
Introduction

During the last decades, the interest for systems of differential equations depending on the past history has been increasing. In fact, the introduction of the delay in the models allows a better description of the real phenomena and a more reliable prediction of their behavior. Such delay models, also called systems with memory or aftereffect, hereditary or time delay systems, are mathematically described by Retarded Functional Differential Equations (RFDEs). Their dynamics is significantly influenced by the presence of the delay terms and oscillations, instability, chaos and loss of performance as well as improved stability can occur. The reason for this more complex dynamics is that, opposite to Ordinary Differential Equations (ODEs), RFDEs are infinite dimensional dynamical systems. "Krasovskii [123] was the first to emphasize the importance of considering the state of a system defined by a functional differential equation" as a function [91, p. 74].

Several monographs have been written on RFDEs and the theory is well-developed [12, 22, 70, 72, 91, 93, 121, 122, 126]. Nowadays it is also recognized their important role in different applied fields, as manifested by the numerous books dealing with applications and numerical methods [18, 20, 81, 93, 103, 121, 125, 126, 147, 176, 178]. Let us soon say that in the present book, we adopt the simpler terminology Delay Differential Equations (DDEs), widely used in applications, instead of RFDEs, more rigorously used in mathematics.

Our main interest is on the stability issue, which is the first relevant task from the point of view of dynamical systems. Having in mind the Principle of Linearized Stability, which reduces the study of the stability of a particular solution of a nonlinear system to the study of the linearized version, the focus is on the stability of linear DDEs and on the numerical methods recently developed to this aim by the authors and published in the main references [38, 44].

In this introduction, we enter into the subject by starting from ODEs, which represent finite dimensional dynamical systems. This allows us to easily introduce in a "colloquial" style basic concepts and classic definitions, but also to emphasize the theoretical and numerical obstacles to be overcome when trying to extend to DDEs. We are moving indeed from finite to infinite dimensional dynamical systems and new challenges arise. This path represents a quite natural approach to introduce the topic under consideration and to give the motivations and, in fact, is followed by other authors as well. Among them, let us mention the nice and complete introduction of the recent book [106], which certainly inspired us.

This introductory chapter starts in Sect. 1.1 by summarizing very basic and classic results on linear ODEs and on the stability of their zero solution. We try to extend the same concepts and techniques to DDEs by examples, encountering several problems in this direction. Besides the Hayes equation and the Cushing equation, both treated in Sect. 1.2, the examples in Sect. 1.3 and Sect. 1.4 are suggested by populations dynamics and mechanical engineering, respectively. Then, Sect. 1.5 resumes from the above difficulties motivating the contents of the book, a complete overview of which is eventually given.

1.1 Linear Ordinary Differential Equations

The scalar linear autonomous ODE

$$x'(t) = ax(t) \tag{1.1}$$

for $a \in \mathbb{R}$ has the unique solution $x(t) = e^{at}u$ for some constant $u \in \mathbb{R}$ depending on the value prescribed at a given time instant. As $t \to \infty$, this solution either decays to zero if $a < 0$ or grows indefinitely if $a > 0$. The case $a = 0$ trivially gives constant solutions.

Dealing with the linear autonomous system of ODEs

$$x'(t) = Ax(t) \tag{1.2}$$

for $A \in \mathbb{R}^{d \times d}$ is not much more difficult. Indeed, looking for a nontrivial exponential solution $x(t) = e^{\lambda t}u$, $u \in \mathbb{R}^d$, leads to the so-called *characteristic equation*

$$\det(\lambda I_d - A) = 0,$$

where I_d is the identity matrix in $\mathbb{R}^d$. Its solutions $\lambda \in \mathbb{C}$, known as *characteristic roots* (or *exponents*), are the eigenvalues of the matrix A. If A has d linear independent eigenvectors, it is easy to show that any solution of (1.2) is a linear combination of exponential functions with the characteristic roots as exponents. Therefore, in the same spirit of the scalar case, the long-time behavior of a solution is determined by the sign of the real part of the characteristic roots. In particular, the essential information resides in the root with the greatest real part, i.e., the *rightmost* one in the complex plane. Indeed, if the latter is to the left of the imaginary axis then the solution decays to zero as $t \to \infty$. Otherwise, if it is to the right then the solution grows indefinitely. If A does not possess d independent eigenvectors, the same conclusions can be obtained through the Jordan canonical form of A (for the latter see, e.g., [73, 145]).

1.1 Linear Ordinary Differential Equations

We thus see that for linear autonomous ODEs, all the solutions go to zero asymptotically and independently of u if (and only if) the rightmost characteristic root has negative real part. Then we talk about *global asymptotic stability* of the zero solution of (1.2). If it has positive real part, all the solutions grow unbounded and we talk about *instability*. In addition, when (1.2) is the result of the *linearization* of a nonlinear autonomous system at a specific constant solution (i.e., an *equilibrium*), the fact that the characteristic roots of the linearized system lie in the open left-half plane guarantees the *local* asymptotic stability of this solution. The latter is a consequence of the celebrated Principle of Linearized Stability.

All the above are classic results in the theory of ODEs and of the associated dynamical systems, contained in almost any introductory monograph on the subject, see, e.g., [61, 183].

Let us also mention that, from the dynamical systems point of view, rightmost characteristic roots crossing the imaginary axis due to the variation of parameters give rise to *bifurcations*, i.e., qualitative changes in the behavior of the solutions. For a complete treatment of the theory of bifurcations see, e.g., [127]. A more introductory monograph is [92]. The simplest instance of bifurcation is (1.1), whose trivial solution is asymptotically stable when $a < 0$ and unstable when $a > 0$. Indeed, a is the only characteristic root (eigenvalue of itself being scalar), it is real, and it can cross the imaginary axis only through the origin, i.e., by changing sign.

A step forward w.r.t. (1.2) is the linear nonautonomous system

$$x'(t) = A(t)x(t) \tag{1.3}$$

for $A : t \mapsto A(t) \in \mathbb{R}^{d \times d}$. Of great interest is the periodic case, i.e., when $A(t) = A(t + \omega)$ for any t and for some minimal period $\omega > 0$. The stability of the zero solution follows from the classic Floquet theory (see the original paper [84], the monograph [82] or, again, [61, 183]). Briefly, if $\Phi(t)$ is any fundamental matrix solution of (1.3), then there exists a periodic matrix $P(t) = P(t + \omega)$ normalized as $P(0) = I_d$ and a constant (possibly complex) matrix C such that $\Phi(t) = P(t)e^{Ct}$ for any t. In particular, $\Phi(\omega) = e^{C\omega}$ is called the *monodromy matrix* and its eigenvalues the *characteristic multipliers*, i.e., the values $\mu \in \mathbb{C}$ satisfying

$$\det(\mu I_d - \Phi(\omega)) = 0.$$

Then Floquet main theorem states that the zero solution of (1.3) is asymptotically stable if and only if the characteristic multipliers lie strictly inside the unit circle in $\mathbb{C}$, unstable when some fall outside. The same criterion can be transferred to the matrix C, whose eigenvalues λ are called again characteristic roots or exponents (or Floquet exponents). Note, in fact, that if μ is a characteristic multiplier then any λ such that $\mu = e^{\lambda \omega}$ is a characteristic root and vice-versa (observe that $\text{Im}(\lambda)$ does not affect the stability). Alternatively, one can directly reduce (1.3) to the autonomous ODE $y'(t) = Cy(t)$ by the change of coordinates $y(t) = P^{-1}(t)x(t)$. Therefore, one can also look to the characteristic equation

$$\det(\lambda I_d - C) = 0.$$

The real difficulty is that, in general, there is no explicit form for C or $\Phi(\omega)$. This is a main disadvantage w.r.t. the autonomous case (where A is available), which persists in the case of DDEs. Anyway, a series of numerical methods can be applied for obtaining suitable approximations to the roots or to the multipliers, see, e.g. [106, Sect. 1.2] and the references therein.

Moreover, similarly to what mentioned above for the autonomous case, bifurcations take place when the *dominant* multiplier, i.e., the one with the largest modulus, crosses the unit circle. We again refer to [127] for a complete treatment of bifurcations of dynamical systems arising also from linear periodic ODEs.

Let us also observe that (1.3) can be either originally linear periodic, in which case the stability information has a global character, or resulting from the linearization of a nonlinear system at a specific periodic solution, in which case the stability information on that specific solution is local (again through the Principle of Linearized Stability).

Eventually, let us give a brief comment on the use of the characteristic equation. Either in the autonomous and in the periodic case, it corresponds to a standard eigenvalue problem for matrices, for which successful and accomplished numerical methods are available (for a general reference see, e.g., [86]). Also efficient routines are part of the standard MATLAB package, as `eig.m` [4] and `eigs.m` [5]. One can also decide to attack directly the characteristic equation with root-finding techniques (e.g., Newton-like methods, see [69]), since it corresponds to compute the roots of a polynomial. If this polynomial is known analytically, which is almost always the case of very low dimensional systems, its numerical solution is safe enough. But if the polynomial coefficients have to be numerically computed starting from the model coefficients (through the determinant), unavoidable rounding errors can cause inaccurate results (see the celebrated *perfidious polynomial* of Wilkinson [201, 202]).

1.2 Simple Linear Delay Differential Equations

Let us consider the scalar DDE with a single constant delay $\tau > 0$

$$x'(t) = ax(t) + bx(t - \tau) \tag{1.4}$$

for $a, b \in \mathbb{R}$.

Being among the simplest examples of DDEs, it is widely taken as a prototype model for several studies regarding either the analytical aspects such as the stability of the zero solution (see, e.g., the monographs [22, 70, 91, 93, 106, 122, 147, 176, 178]) or the numerical aspects of relevant time-integrators such as stability and convergence (see, e.g., the monographs [20, 90]).

Nevertheless, the condition on the parameters a and b guaranteeing the stability of the zero solution were first investigated in [95] and, therefore, (1.4) is also cited as

1.2 Simple Linear Delay Differential Equations

the Hayes equation. Such conditions are recovered by looking at the characteristic equation

$$\lambda - a - be^{-\lambda\tau} = 0, \qquad (1.5)$$

obtained by seeking for a nontrivial exponential solution $x(t) = e^{\lambda t} u$, $u \neq 0$, of (1.4), exactly as done for ODEs in the previous section. Indeed, the solutions $\lambda \in \mathbb{C}$ of (1.5) are still called characteristic roots. Moreover, extending from ODEs, it is a classic result that the zero solution of (1.4) is asymptotically stable if and only if $\text{Re}(\lambda) < 0$ for all these roots, whereas it is unstable as soon as one of them has positive real part, see, e.g., [70, Chap. I, Corollary 5.5] or [91, Chap. 7, Theorem 4.1]. The fundamental difference (and difficulty) w.r.t. the case of ODEs is that DDEs have infinitely many characteristic roots, given that the characteristic equation is transcendental or, specifically, a quasi-polynomial equation [91, Appendix].

The study of the location of the characteristic roots of (1.4), first appeared in [22, 95], originated from a more general branch of research devoted to study the zeros of exponential polynomials, see, e.g., [129, 159, 169, 182, 200, 203]. Indeed, as claimed in [129], first attempts to "focus the attention in large measure upon the geometric determination of the configuration of regions in which the roots of the equation are located" can be found in [159, 169].

The analysis of the stability of the zero solution of (1.4) w.r.t. the parameters a and b is nowadays largely known, and the resulting picture showing the (a, b)-plane divided into stable and unstable regions, a so-called *stability chart*, can be found in several publications, see, e.g., [19, Figs. 1 and 2], [33, Fig. 2], [70, Fig. XI.1], [71, Fig. 1], [91, Fig. 5.1], [106, Fig. 2.1] and [164, Fig. 1], to name a few. We reproduce in Fig. 1.1 a modified version of the one appearing in [27].

We do not intend here to cover again the necessary steps to get to Fig. 1.1. Indeed, a complete and detailed presentation can be found, e.g., in [70, 106] or in [33], where also the case of complex parameters is tackled. We just summarize that it is enough to set $\lambda = \alpha + i\beta$, $\alpha, \beta \in \mathbb{R}$, in (1.5) and to consider the separated real equations for the real and imaginary parts:

$$\begin{cases} \alpha - a = be^{-\alpha\tau} \cos(\beta\tau), \\ -\beta = be^{-\alpha\tau} \sin(\beta\tau). \end{cases} \qquad (1.6)$$

This allows the exact determination of all the curves drawn in Fig. 1.1 and it represents an elegant application of the classic *D-subdivision method* originating from [160, 150]. In particular, the red curves separate the stability domain ($\alpha < 0$, green colored), from the instability domain ($\alpha > 0$, red colored). Along the blue curve there is a double real root: above there are real roots (one for $b > 0$, two for $b < 0$), below there are only complex-conjugate pairs. The black curves account for a pair of complex-conjugate roots crossing the imaginary axis. In each portion enclosed by all these curves and the dashed line $a + b = 0$ there is a constant number of roots with positive real part, number that can be easily determined by observing that the solutions of (1.5) vary continuously w.r.t. the parameters (compare with [106, Fig. 2.1] or with [33, Fig. 2]). Alternatively, the number of unstable roots can be

Fig. 1.1 Stability chart and characteristic roots of the Hayes equation (1.4)

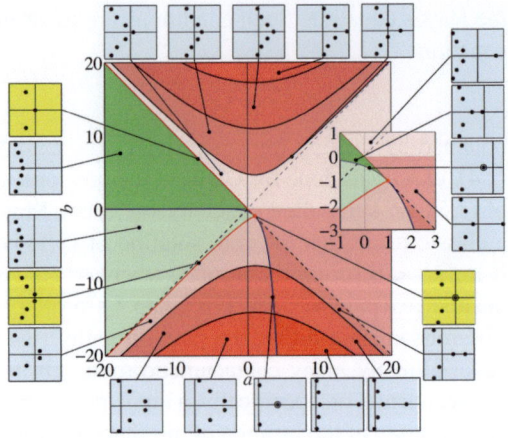

found by using Stepán's formulae [178], as done in [106]. Finally, it is important to observe that in the region $|b| < |a|$ for $a < 0$ the zero solution is asymptotically stable independently of the value τ of the delay.

The D-subdivision method can still be applied in the case of a scalar DDE with a single distributed delay, such as the Cushing equation

$$x'(t) = ax(t) + b \int_{-\tau}^{0} x(t+\theta) d\theta, \tag{1.7}$$

first considered in [66] in the context of population dynamics, or variants with different integral kernels [14, 25, 26, 148]. The relevant stability chart can be found, e.g., in [106, Fig. 2.2].

However, and unfortunately, such a complete analysis is unattainable for general DDEs. Already for the system

$$x'(t) = Ax(t) + Bx(t-\tau), \tag{1.8}$$

with A, B general matrices in $\mathbb{R}^{2\times 2}$, the study of the stability of the zero solution through the analysis of the exact position in $\mathbb{C}$ of its characteristic roots w.r.t. given A and B is still an open problem. Indeed, such barrier is properly stressed either in [91, p.109]: "the exact region of stability as an explicit function of A, B and $[\tau]$ is not known and probably will never be known. The reason is simple to understand because the characteristic equation...is so complicated. It is therefore, worthwhile to obtain methods for determining approximations to the region of stability.", in [70, p. 305]: "...but for characteristic equations in general, the rule seems to be that they are not at all amenable to analysis." or, more recently, in [162, p.1671]: "...checking the eigenvalue conditions is much harder than for [Ordinary Differential Equations]."

1.2 Simple Linear Delay Differential Equations

It starts then to become clear the necessity of suitable numerical strategies to tackle the problem of the stability analysis through the characteristic roots. In the next Section, we give an example of a linear autonomous DDE coming from a linearization procedure. Instead, in Sect. 1.4, we present a nonautonomous periodic equation, which requires a different treatment.

1.3 An Example from Population Dynamics

The analysis of the stability of the zero solution of linear DDEs is fundamental in investigating the local stability of equilibria of nonlinear problems through linearization. The theoretical foundation is provided again by the Principle of Linearized Stability. The basic result holding for ODEs, see, e.g., [92, Theorems 9.5 and 9.7] or [180, Theorem 2.3.5], can be indeed generalized to DDEs, see, e.g., [70, Chap. VII, Corollary 5.12].

As an example, let us consider the (nonlinear) delay logistic equation

$$x'(t) = rx(t)(1 - x(t-1)). \tag{1.9}$$

It models the dynamics of a population with growth rate $r > 0$ and normalized carrying capacity, where competition takes effect one unit of time later on. This model was first introduced in [98], justifying the common name of Hutchinson equation. It represents a further step toward real-life problems w.r.t. previous models such as that of Verhulst [192], who also coined the term *logistic* [193], or the simple exponential model of Malthus [142] (both ODEs).

It is trivial to observe that (1.9) has the trivial equilibrium $\bar{x}_0 = 0$ and the positive equilibrium $\bar{x}_1 = 1$, independently of the parameter r. The linear variational equation at an equilibrium $\bar{x}$ reads

$$x'(t) = r(1 - \bar{x})x(t) - r\bar{x}x(t-1). \tag{1.10}$$

Let us note that the linearization technique is the same as for ODEs, i.e., look for a solution of the form $\bar{x} + x(t)$ with $x(t)$ a perturbation small enough to cancel $o(x(t))$ terms (anyway see also Example 3.1 in Chap. 3). As a consequence of (1.10), it is easy to observe that $\bar{x} = \bar{x}_0$ is always unstable since (1.10) becomes

$$x'(t) = rx(t)$$

and $r > 0$. On the other hand, for $\bar{x}_1 = 1$ we obtain

$$x'(t) = -rx(t-1), \tag{1.11}$$

which is a pure DDE and, in particular, a special case of the Hayes equation (1.4). Indeed, one can recover the local stability properties of the positive equilibrium as r

varies just by looking at Fig. 1.1 for $a=0$ and $b=-r<0$, i.e., along the vertical downward semiaxis. This would be sufficient to conclude that $\bar{x}_1$ is asymptotically stable for $r<r^*$ with $r^*:=\pi/2$, value at which a Hopf bifurcation [92, 127] occurs with a stable periodic solution arising for r slightly above r^*. The latter fact can be proven rigorously by following the same arguments of [91, Sect. 11.4].

As a consequence, let us note that, diversely from (nonlinear) scalar ODEs that can exhibit only exponential (and constant) behavior, (nonlinear) scalar DDEs can have periodic solutions due to the presence of complex-conjugate pairs of characteristic roots (of the linearized system). The dynamics is thus more rich and, in fact, also chaotic motion can manifest as it is the case of the celebrated Mackey-Glass equation [138] (Example 3.1).

We show here that the same conclusion can be obtained algebraically by following a different path. A detailed exposition can be found in [33]. The characteristic equation associated to (1.11) reads

$$\lambda + re^{-\lambda} = 0.$$

It can be decomposed into the two real equations

$$\begin{cases} \alpha = -re^{-\alpha}\cos(\beta), \\ \beta = re^{-\alpha}\sin(\beta), \end{cases} \quad (1.12)$$

once $\lambda = \alpha + i\beta$, $\alpha, \beta \in \mathbb{R}$, is substituted.

The search for real roots ($\beta = 0$) is easy: they are given, if any, by the solutions of $-\alpha = re^{-\alpha}$. Therefore, there are no real roots for $r > 1/e$, there is one double real root $\lambda = -1$ for $r = 1/e$, and there are two distinct real roots $\lambda_1 < -1$ and $\lambda_2 \in (-1, 0)$ for $r < 1/e$. It is also immediate to verify that $\lambda_1 \to -\infty$ and $\lambda_2 \to 0^-$ as $r \to 0^+$. One concludes that real roots, whether existing, cannot cause instability: the nontrivial equilibrium $\bar{x} = \bar{x}_1$ may loose stability only in favor of a stable periodic solution, i.e., through a Hopf bifurcation. This is in accordance with what is observed before via Fig. 1.1.

To rigorously check the above finding one has to look for complex-conjugate pairs. Without loss of generality (since r is real), let us assume $\beta > 0$. By taking the ratio member-to-member of the two equations in (1.12) one finds that $\lambda = \alpha + i\beta$ is a characteristic root if and only if the point (β, α) belongs to the graph of the function

$$\alpha(\beta) = -\frac{\beta}{\tan(\beta)}, \quad \beta \neq k\pi, \quad k = 1, 2, \ldots, \quad (1.13)$$

in the (β, α)-plane. On the other hand, by squaring and summing member-to-member the same two equations, one similarly concludes that $\lambda = \alpha + i\beta$ is a characteristic root if and only if the point (α, β) belongs to the graph of the function

$$\beta(\alpha) = \sqrt{r^2 e^{-2\alpha} - \alpha^2}, \quad |\alpha| < re^{-\alpha}, \quad (1.14)$$

in the (α, β)-plane.

1.3 An Example from Population Dynamics

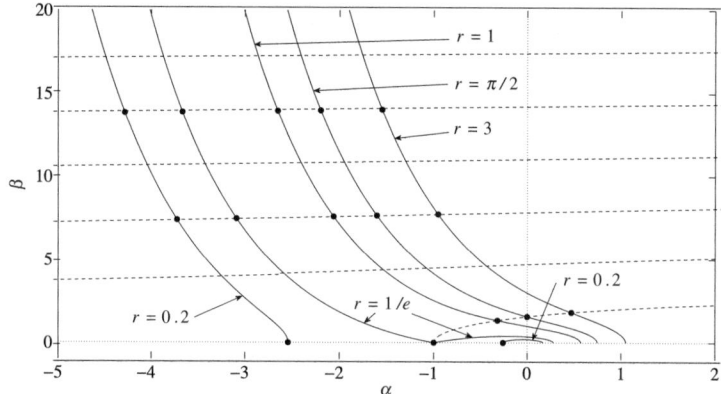

Fig. 1.2 Characteristic roots (•) of (1.11) as intersections of the graphs of (1.13) and (1.14) for varying r

Now, the function (1.13) is not globally invertible, but every branch between $(k\pi, (k+1)\pi)$, $k = 0, 1, 2, \ldots$, is so. Therefore, one can transfer it to the same (α, β)-plane of (1.14) and look for the characteristic roots as the intersections of the two graphs. Among all such intersections, half have to be discharged due to the squaring procedure adopted in obtaining (1.14). It is easy to check that these spurious intersections are those in the range $\beta \in (k\pi, (k+1)\pi)$ for k odd. These graphs are shown in Fig. 1.2 for various values of r: dashed lines for (1.13) and solid lines for (1.14). Since (1.13) is independent of r, an increasing of the latter provokes a rightward movement of the graph of (1.14) as well as of its intersections with the graph of (1.13). This lets us conclude that the rightmost complex-conjugate pair crosses the imaginary axis when $\beta = \pi/2$ (the first zero of (1.13)) and this happens in fact when $r = r^*$ (from the second of (1.12) for $\beta = \pi/2$). Therefore, the expected Hopf bifurcation is confirmed.

Again, unfortunately, the possibility to extend this approach to more general DDEs is limited. The scalar case of the Hayes equation (1.4) with complex coefficients is still amenable of such analysis [33]. Attempts can be made to study the case with two constant delays, but already the Cushing equation (1.7) is prohibitive. As already remarked, the case of systems, such as, e.g., (1.8), is even more complicated. In fact, the characteristic equation, as for systems of ODEs, is given through the determinant of a matrix, so that it is already difficult to write down explicit equations like (1.6) or (1.12).

1.4 An Example from Mechanical Engineering

Delay is ubiquitous in engineering applications. As a starting reference see [81, 106]. The Mathieu equation with delay is a well-known prototype model for Newtonian problems with both delay and periodic coefficients.

The ordinary Mathieu equation

$$x''(t) + a_1 x'(t) + (\delta + \varepsilon \cos(t))x(t) = 0$$

was originally considered in [144] in the study of the vibration of an elliptic membrane. Other instances appeared in dealing with a pendulum oscillating under parametric forcing [118, 130, 179].

On the other hand, the delayed oscillator

$$x''(t) + a_1 x'(t) + \delta x(t) = b x(t - \tau)$$

has gained an increasing interest since the publication of the relevant stability chart in [97], becoming a classic in Newtonian problems with delay [54, 63, 122, 126, 139, 143, 171, 178].

The combination of the two effects, namely delay and parametric forcing, results in the class of delayed Mathieu equations, which has been extensively considered in the literature. The recent monograph [106], from which the following example is taken, collects a number of stability investigations on the subject, performed through the semi-discretization method of the authors. A complete bibliography can be found therein. Also, in the last decade, the delayed Mathieu equation has often been taken as a benchmark to test different techniques (either analytical and numerical) in order to gain insight into the analysis of its stability properties, especially when the parameters of the model are uncertain or varying, see, e.g., [39, 51, 53, 120, 44].

We consider here the mechanical model of stick-balancing with parametric excitation described in [106, Sect. 5.4] and first treated in [102]. It represents a common example of the use of parametric forcing to control unstable dynamics, see again [118, 130, 179], but see also [59, 137] for other related applications in engineering.

The model consists of a stick attached to a horizontal slide mounted on a base which moves periodically up and down, Fig. 1.3. The stick is assumed to be homogeneous with mass m and length l, the mass of the slide is assumed to be negligible w.r.t. m and the slide base moves according to $r \cos(\Omega t)$. The angular position φ of the stick and the horizontal position x of the pivot point in the slide are considered as general coordinates. A feedback force Q is applied to the slide to balance the stick, trying to keep it in the vertical upright position. Such force is assumed to depend on both the angular position φ and the angular velocity φ' of the stick. A delay τ enters the model naturally due to this feedback control. The equations of motion turn out to be

1.4 An Example from Mechanical Engineering

Fig. 1.3 Stick-balancing with parametric forcing

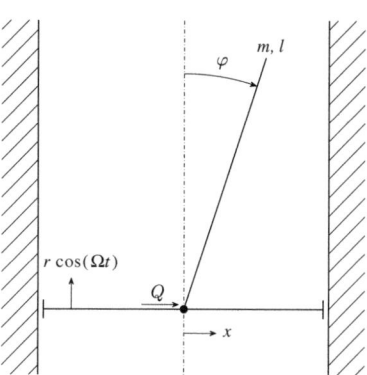

$$\begin{cases} \dfrac{1}{3}ml^2\varphi''(t) + \dfrac{1}{2}ml\cos(\varphi(t))x''(t) \\ \quad + \left(-\dfrac{1}{2}mgl + \dfrac{1}{2}mlr\Omega^2\cos(\Omega t)\right)\sin(\varphi(t)) = 0, \\ \dfrac{1}{2}ml\cos(\varphi(t))\varphi''(t) + mx''(t) - \dfrac{1}{2}ml(\varphi'(t))^2\sin(\varphi(t)) = Q(\varphi(t-\tau),\varphi'(t-\tau)). \end{cases}$$

The horizontal displacement x can be eliminated, leading to the single equation for φ

$$\left(\dfrac{1}{3}ml^2 - \dfrac{1}{4}ml^2\cos^2(\varphi(t))\right)\varphi''(t) + \dfrac{1}{8}ml^2(\varphi'(t))^2\sin(2\varphi(t))$$
$$+ \left(-\dfrac{1}{2}mgl + \dfrac{1}{2}mlr\Omega^2\cos(\Omega t)\right)\sin(\varphi(t))$$
$$= -\dfrac{1}{2}lQ(\varphi(t-\tau),\varphi'(t-\tau))\cos(\varphi(t)),$$

which is a nonlinear second-order DDE with a single constant delay.

Now, by assuming a force (locally) linear w.r.t. both position φ and velocity φ', linearization around the upright position $\bar{\varphi} = 0$ leads to

$$\dfrac{1}{12}ml^2\varphi''(t) + \left(-\dfrac{1}{2}mgl + \dfrac{1}{2}mlr\Omega^2\cos(\Omega t)\right)\varphi(t)$$
$$= -\dfrac{1}{2}l\left(K_p\varphi(t-\tau) + K_d\varphi'(t-\tau)\right).$$

A rearrangement of the parameters finally gives the delayed Mathieu equation

$$\varphi''(t) + (\delta + \varepsilon\cos(\Omega t))\varphi(t) = b_1\varphi(t-\tau) + b_2\varphi'(t-\tau). \qquad (1.15)$$

The analysis of the stability of the zero solution is based on the original Floquet Theory for ODEs as extended to DDEs, see [70, Chap. XIII] or [91, Chap. 8]. The subject is summarized in Chap. 4. As for ODEs, this analysis is based on the knowledge of the associated characteristic multipliers and, in particular, of their position w.r.t. the unit circle. This knowledge is, in general, not attainable analytically even because, as for the autonomous case, there are infinitely many multipliers.

A more general class including (1.15) is analyzed in Sect. 8.3 by the numerical method proposed in Chap. 6.

1.5 Scopes and Synopsis

From the previous sections, one can infer that basic concepts (characteristic equations, roots and multipliers, stability) and techniques (linearization, bifurcation analysis) borrowed from the theory of ODEs can be suitably extended to DDEs. The price to pay is that of an increasing difficulty, stressed, e.g., by the following facts:

- a more rich dynamics already in the scalar case, i.e., periodic and chaotic motions;
- characteristic equations are transcendental, difficult to write explicitly and with infinitely many solutions.

As a consequence of the latter, it seems unattainable to analyze the stability properties of the zero solution of linear DDEs by solving the characteristic equation, i.e., by computing the characteristic roots in the autonomous case or the characteristic multipliers in the periodic one. And indeed this approach works only for basic and very simple examples of DDEs.

Nevertheless, it is sufficient to change the point of view to gain a deeper insight into the problem. In fact, it will be clear that roots and multipliers can be seen as eigenvalues of suitable infinite dimensional linear operators. These operators play the role of the matrix A in (1.2) for linear autonomous DDEs or that of the monodromy matrix $\Phi(\omega)$ associated to (1.3) for linear periodic DDEs. The remarkable difference at the core is that DDEs generate dynamical systems on infinite dimensional state spaces whereas ODEs generate dynamical systems on finite dimensional state spaces.

Beyond being an elegant and powerful outcome of the operatorial approach, this alternative characterization paves the way to compute roots and multipliers as the eigenvalues of these operators. Still we are left with problems in infinite dimension, that is why we resort to numerical analysis in order to approximate (a finite number of) these eigenvalues in a possibly accurate and effective manner. This is the main scope motivating the content of the present book, whose synopsis follows.

In Part I, we acquire the necessary theoretical background. In particular, in Chap. 2, we recall basic results such as existence, uniqueness, continuous dependence and stability of solutions of DDEs, beyond defining the general class of linear equations we aim at considering. With Chap. 3 we enter the dynamical systems point of view for the linear autonomous case. We illustrate the functional analytic framework based on the semigroup of solution operators, its infinitesimal generator and their spectral

1.5 Scopes and Synopsis

properties, intimately related to the characteristic roots. In parallel, Chap. 4 summarizes the extension of the Floquet theory to linear periodic DDEs. This is done in the more ample context of general nonautonomous problems (i.e., not necessarily periodic), by introducing the family of evolution operators. The monodromy operator for periodic DDEs is a particular instance and the characteristic multipliers are related to the spectrum of the latter.

Part II deals with the numerical analysis. Chapter 5 presents in detail the discretization of the infinitesimal generator of linear autonomous DDEs with the pseudospectral differentiation method. It analyzes the algorithm as well as the convergence of the approximated eigenvalues to the exact characteristic roots. Chapter 6 discusses the discretization of the evolution operator for linear nonautonomous DDEs. It constructs the algorithm based on the pseudospectral collocation method and analyzes the convergence of the spectral elements. As particular instances, the resulting method can be applied either to approximate the characteristic multipliers of linear periodic DDEs or the characteristic roots of linear autonomous DDEs.

The last part of the book, Part III, is divided into Chaps. 7 and 8. In the former, we show how the algorithms of Part II are implemented in MATLAB. In the latter we present, first, a series of case studies to test the performance of the MATLAB codes and, second, a couple of real-life applications. The scope is that of guiding the interested readers in properly using the codes on benchmark examples so to make them autonomous with their own models.

Let us note that while Part I is nowadays a standard expertise of theoretical results and tools in the field of DDEs, the numerical methods presented in Part II and implemented and applied in Part III represent the outcome of the research of the authors of the last 15 years or so. The two main reference papers, including proofs of convergence, are [38] for the discretization of the infinitesimal generator with the pseudospectral differentiation method and [44] for the discretization of the solution operator with the pseudospectral collocation method.

The approximation of the eigenvalues of the solution operator first appeared in [80] for linear autonomous DDEs. There the authors used Linear Multistep methods and the proposed algorithm was originally used in the MATLAB package DDE-BIFTOOL for the bifurcation analysis of DDEs [78, 79]. A more efficient implementation has been reached with [191]. From [80], the idea of determining the stability of linear autonomous DDEs through the computation of the eigenvalues of either the solution operator or the infinitesimal generator has been deeply investigated in [30], where for both the approaches Linear Multistep, Runge-Kutta, and pseudospectral methods have been analyzed (see also [29]). For the approaches based on Runge-Kutta methods see in particular [31, 37]. For a general reference on Linear Multistep and Runge-Kutta methods see, e.g., [90, 128], for spectral and pseudospectral methods see, e.g., [57, 87, 184].

The efficacy demonstrated by pseudospectral methods motivated also the extension of the approach based on the infinitesimal generator to more general classes of functional differential equations [34–36, 40, 42, 43, 45, 155, 187]. Moreover, the approach revealed particularly useful in studying epidemics and more general population dynamics [23, 50, 46, 85, 133].

In the context of linear autonomous DDEs, other methodologies exist to detect stability by way of the rightmost characteristic roots. Among these, let us mention Galerkin projection [197, 198], the Cluster Treatment of Characteristic Roots [153, 154, 172–174], the mapping procedure [194–196], the harmonic balance [131], the method of steps [117], and the approach [206, 207] based on the Lambert W function [64, 111]. As a general reference, see also [109]. Eventually, methods were developed to solve efficiently the characteristic equation seen as a nonlinear eigenvalue problem [110, 112–116, 146 189].

As far as periodic DDEs are concerned, the idea of discretizing the monodromy operator based on truncating a spectral expansion via Chebyshev polynomials first appeared in [54], later refined in [51, 53]. Independently, pseudospectral collocation first appeared in [39], with a complete convergence analysis in the above cited paper [44]. Other discretization techniques such as finite or spectral elements are used in [16, 119, 120, 143]. An approach based on characteristic matrices is presented in [170, 181], while Galerkin projection is used in [9].

A particular mention is reserved to the semi-discretization method of Insperger and Stepán, an efficient scheme to produce stability charts in the parameters domain. A complete treatment appears in the already cited monograph [106], for the original papers see [103, 105, 107].

Let us cite also the monograph [147] for an extensive treatment of some of the above cited approaches based on the computation of eigenvalues. Instead, it is worthy underlying that, of course, resorting to characteristic roots and multipliers is not the unique way to determine stability of linear DDEs. The monograph [151] can serve as a general reference for alternative techniques such as Lyapunov methods or Linear Matrix Inequalities. Note that the latter furnish only sufficient yet not necessary conditions.

Eventually, let us go back to the comment on the use of the characteristic equation given at the end of Sect. 1.1 for ODEs. Similar arguments extend to the infinite dimensional case of DDEs. An example can be found in [30].

Part I
Theory

After the description of an evolution phenomenon by a mathematical model, the next step in the study of the dynamics is to set the mathematical framework, which inserts the specific instance into a more abstract context. The collection of definitions and theorems traces the "boundaries" of the theory: it allows to determine the applicability and the features of the mathematical model, to lay the foundation for the construction of numerical models and, finally, to validate the results of the numerical simulations. The purpose of this theoretical part is to introduce the basic notation and to present some background material, required throughout the book. Nowadays, the list of books dealing with theory and applications of DDEs is quite long [12, 20, 22, 70, 72, 81, 91, 93, 121, 122, 126, 147, 151, 176]. In particular, we mainly follow [70, 91, 121], where the interested reader can find further details.

Before studying the dynamics, one wants first to be sure that the mathematical problem is well-posed, i.e., there exists a unique solution, which depends continuously on the data (original definition in [89]). Therefore, in the first chapter, after setting the basic notation, we recall some classic results on the solvability of the Cauchy problem for DDEs. In the applications, it is important to identify some particular solutions, such as, e.g., the equilibria, and to predict the effect of small perturbations. This concerns the stability theory and the corresponding definitions are given in Sect. 2.3. By the Principle of Linearized Stability, the stability analysis of any solution of interest of a nonlinear equation can be attributed to the behavior of the corresponding linearized DDE. In this context, the understanding of the linear case is crucial, adding a further motivation for its study. In fact, as already pointed out in the introductory examples of Chap. 1, linear DDEs arise when modeling different real-life linear phenomena or by linearization of nonlinear ones.

The focus is on linear DDEs of autonomous and periodic type, whose necessary theory is summarized in Chaps. 3 and 4, respectively. Such classes of linear equations are of great interest in applications and the stability theory is well established. We remark that the linearization of nonlinear autonomous DDEs at equilibria and periodic solutions leads, respectively, to linear autonomous DDEs and linear periodic DDEs.

For completeness, we recall that the method of Lyapunov functionals has been successfully applied to examine the stability of solutions of nonlinear DDEs [91, 121, 122]. The approach furnishes sufficient conditions for stability and instability, generalizing the method of Lyapunov for ODEs.

Chapter 2
Notation and Basics

The aim of this chapter is to introduce basic notation and definitions, together with solvability theorems for Cauchy problems for DDEs and a remark on continuous dependence on the data. It is a preparatory work not only for the next theoretical chapters, but also for the numerical approaches, core of this monograph, presented in Part II and Part III. Finally, we introduce the definitions of stability of a given solution and the Principle of Linearized Stability in its generality. We refer to [70, 91, 121] for further details and for the proofs which are not given.

2.1 Notation

We denote the independent variable *time* by t, $t \in \mathbb{R}$, the *dimension* of the system (i.e., the number of equations) by d, $d \in \mathbb{N}$ with $d \geq 1$, and the dependent variable by the map $x : t \mapsto x(t) \in \mathbb{R}^d$, $x = (x_1, \ldots, x_d)^T$. Let $\tau > 0$ be the *maximum delay* of the system. We denote by X the *state space* of continuous functions $C([-\tau, 0], \mathbb{R}^d)$, which is a Banach space when, as we choose here, it is equipped with the maximum norm $\|\varphi\|_X = \max_{\theta \in [-\tau, 0]} \|\varphi(\theta)\|_\infty$, $\varphi \in X$, where $\|\cdot\|_\infty$ is the infinity norm on $\mathbb{R}^d$. The relevant matrix induced norm, as well as induced norms for operators, are denoted simply by $\|\cdot\|$, unless differently specified.

We consider *nonlinear nonautonomous* DDEs of the form

$$x'(t) = F(t, x_t), \quad t \in \mathbb{I}, \tag{2.1}$$

for an interval $\mathbb{I} \subseteq \mathbb{R}$ unbounded on the right and a continuous function $F : \mathbb{I} \times X \to \mathbb{R}^d$. $x_t \in X$ is the *state* at time t defined, according to the standard Hale-Krasovsky notation [123], as

$$x_t(\theta) := x(t + \theta), \quad \theta \in [-\tau, 0]. \tag{2.2}$$

Notice that for DDEs the symbol $'$ denotes the right-hand derivative [91, p. 36]. When F does not depend explicitly on time, then $\mathbb{I} = \mathbb{R}$ and we call *autonomous* the resulting DDE:

$$x'(t) = F(x_t), \; t \in \mathbb{R}. \tag{2.3}$$

Linear nonautonomous DDEs are described by

$$x'(t) = L(t)x_t, \; t \in \mathbb{I}, \tag{2.4}$$

where $L(t) : X \to \mathbb{R}^d$, $t \in \mathbb{I}$, is a linear and bounded functional and the map $L(\cdot)\psi : \mathbb{I} \to \mathbb{R}^d$, $\psi \in X$, is continuous. L is representable as the Lebesgue–Stieltjes integral

$$L(t)\psi = \int_{-\tau}^{0} d_\theta [\eta(t,\theta)] \psi(\theta), \; t \in \mathbb{I}, \; \psi \in X, \tag{2.5}$$

where $\eta(t, \cdot)$, $t \in \mathbb{I}$, is a Normalized Bounded Variation (NBV) function [165, Riesz Representation Theorem]. By assuming that $t \mapsto \eta(t, \cdot)$ is continuous when NBV$([-\tau, 0], \mathbb{R}^{d \times d})$ is equipped with the total variation norm, the left-hand side of (2.5), i.e., the function $(t, \psi) \mapsto L(t)\psi$, is continuous on $\mathbb{I} \times X$. For a summary introduction on NBV functions and abstract integration see [70, Appendix I].

When there is $\omega > 0$ such that $L(t + \omega) = L(t)$ for all $t \in \mathbb{R}$, then $\mathbb{I} = \mathbb{R}$ and

$$x'(t) = L(t)x_t, \; t \in \mathbb{R}, \tag{2.6}$$

is called *periodic*. When (2.4) is autonomous, i.e., $L = L(t)$ is independent of t, then $\mathbb{I} = \mathbb{R}$ and we simply write

$$x'(t) = Lx_t, \; t \in \mathbb{R}. \tag{2.7}$$

Linear DDEs (2.4) arising in the applications of interest have the form

$$x'(t) = A(t)x(t) + \sum_{k=1}^{p} B_k(t) x(t - \tau_k) + \sum_{k=1}^{p} \int_{-\tau_k}^{-\tau_{k-1}} C_k(t, \theta) x(t+\theta) d\theta, \; t \in \mathbb{I}, \tag{2.8}$$

where $0 < \tau_1 < \cdots < \tau_p := \tau$ are p distinct delays (we also set $\tau_0 := 0$ for convenience), $A : \mathbb{I} \to \mathbb{R}^{d \times d}$, $B_k : \mathbb{I} \to \mathbb{R}^{d \times d}$ and $C_k : \mathbb{I} \times [-\tau_k, -\tau_{k-1}] \to \mathbb{R}^{d \times d}$ for $k = 1, \ldots, p$. The above assumptions on η are fulfilled whenever the functions A and B_k are continuous, $C_k(\cdot, \theta)$ is continuous for all $\theta \in [-\tau_k, -\tau_{k-1}]$ and, for any compact interval I in $\mathbb{I}$, there exists $\gamma_k \in C([-\tau_k, -\tau_{k-1}], \mathbb{R})$ such that $\|C_k(t, \theta)\| \leq \gamma_k(\theta)$ for all $t \in I$ and $\theta \in [-\tau_k, -\tau_{k-1}]$. We call $A(t)x(t)$ the current time term and, for $k = 1, \ldots, p$, $B_k(t)x(t - \tau_k)$ the kth discrete delay term and $\int_{-\tau_k}^{-\tau_{k-1}} C_k(t, \theta) x(t + \theta) d\theta$ the kth distributed delay term. When $p = 1$, we

2.1 Notation

simply have $\tau_1 = \tau$ and write $B_1 = B$ and $C_1 = C$. When (2.8) is autonomous, the dependence on time of A, B_k, and C_k is suppressed. Hereafter, we call (2.8) the *prototype* DDE.

2.2 The Cauchy Problem

This section is concerned with well-posedness of Cauchy problems for DDEs (2.1).

Definition 2.1 (*solution*) Given $t_0 \in \mathbb{I}$, a solution x of (2.1) on $[t_0 - \tau, t_f) \subseteq \mathbb{I}$ is a continuous function $x : [t_0 - \tau, t_f) \to \mathbb{R}^d$ which satisfies (2.1) on $[t_0, t_f)$.

To specify a solution we assign an initial condition, i.e., a function $\varphi \in X$ at a certain initial time t_0. Given $(t_0, \varphi) \in \mathbb{I} \times X$, the Cauchy problem for (2.1) is defined as

$$\begin{cases} x'(t) = F(t, x_t), & t \geq t_0, \\ x(t_0 + \theta) = \varphi(\theta), & \theta \in [-\tau, 0]. \end{cases} \quad (2.9)$$

Definition 2.2 (*solution of the Cauchy problem*) Given $(t_0, \varphi) \in \mathbb{I} \times X$, a solution x of (2.9) on $[t_0 - \tau, t_f) \subseteq \mathbb{I}$ is a solution on $[t_0 - \tau, t_f)$ with initial function φ, i.e., $x_{t_0} = \varphi$. x is called global if it is defined on $[t_0 - \tau, +\infty)$.

To emphasize the dependence on t_0 and φ, we sometimes write $x(t; t_0, \varphi)$ for the solution at time t of (2.9). Various theorems on existence and uniqueness of solutions of (2.9) appear in the literature. The key ingredient is the Lipschitz continuity of F w.r.t. the state, either *locally*, i.e., for every $(t_0, \varphi) \in \mathbb{I} \times X$, there exists a neighborhood $\mathscr{U}$ of (t_0, φ) and a constant $\mathrm{Lip}(F)$ such that $\|F(t, \psi_1) - F(t, \psi_2)\|_\infty \leq \mathrm{Lip}(F)\|\psi_1 - \psi_2\|_X$ for all $(t, \psi_1), (t, \psi_2) \in \mathscr{U}$, or *globally*, i.e., there exists a constant $\mathrm{Lip}(F)$ such that $\|F(t, \psi_1) - F(t, \psi_2)\|_\infty \leq \mathrm{Lip}(F)\|\psi_1 - \psi_2\|_X$ for all $(t, \psi_1), (t, \psi_2) \in \mathbb{I} \times X$.

Theorem 2.1 (local solution of the Cauchy problem) *Let F be locally Lipschitz w.r.t. the state. Then, for every $(t_0, \varphi) \in \mathbb{I} \times X$, there exists $t_f > t_0$ and a unique solution x of (2.9) on $[t_0 - \tau, t_f)$. Moreover, x depends continuously on F, t_0 and φ.*

In general, the continuous dependence studies the effect on the solution of the errors in either F and φ [121, p.41], i.e., replacing F and φ in (2.9) by $\tilde{F}$ and $\tilde{\varphi}$ such that

- for any $t_1 \in [t_0, t_f)$ and $\varepsilon > 0$ there is $\delta > 0$ such that $\|\tilde{\varphi} - \varphi\|_X \leq \delta$ and $\|\tilde{F}(t, \psi) - F(t, \psi)\|_\infty \leq \delta$, $t \in [t_0, t_1]$, $\|\psi - x_t\|_X \leq \varepsilon$;
- $\tilde{F}$ and $\tilde{\varphi}$ satisfy the assumptions of Theorem 2.1, so that the perturbed Cauchy problem has a unique solution $\tilde{x}$ on $[t_0 - \tau, \tilde{t}_f)$;

it implies $\sup_{t \in [t_0, \min\{t_f, \tilde{t}_f\}]} \|\tilde{x}(t) - x(t)\|_\infty \leq \varepsilon$. Observe also that under the assumptions in Theorem 2.1, the solution of (2.9) belongs to $C([t_0 - \tau, t_f), \mathbb{R}^d) \cap C^1([t_0, t_f), \mathbb{R}^d)$. More regularity is ensured by further smoothness of F.

Theorem 2.2 (global solution of the Cauchy problem) *Let F be globally Lipschitz w.r.t. the state. Then, for every $(t_0, \varphi) \in \mathbb{I} \times X$, there exists a unique solution of* (2.9) *on $[t_0 - \tau, +\infty)$.*

For autonomous DDEs (2.3), it is not restrictive to assume $t_0 = 0$ (since $x(t; t_0, \varphi) = x(t + t_0; 0, \varphi)$) and we relax the notation $x(\cdot; 0, \varphi)$ to $x(\cdot; \varphi)$. For $\varphi \in X$, (2.9) reads

$$\begin{cases} x'(t) = F(x_t), \ t \geq 0, \\ x_0 = \varphi. \end{cases} \tag{2.10}$$

2.3 Stability of Solutions

Stability, the crucial question addressed in the book, concerns the effects of small perturbations of φ w.r.t. a solution of interest $\bar{x}(\cdot; t_0, \varphi)$ given on $[t_0 - \tau, +\infty)$ under the hypothesis of Theorem 2.2. By considering $y(t) = x(t; t_0, \psi) - \bar{x}(t; t_0, \varphi)$ and the DDE $y'(t) = F(t, y_t + \bar{x}_t) - F(t, \bar{x}_t)$ corresponding to the zero initial function, the definitions of stability and hence the stability analysis of $\bar{x}$ reduce to the stability of the zero solution. Here we prefer to give all the stability definitions directly for $\bar{x}$.

Definition 2.3 (*stable/unstable solution*) The solution $\bar{x}(\cdot; t_0, \varphi)$ of (2.9) is called stable if for any $\varepsilon > 0$ there exists $\delta = \delta(t_0, \varepsilon) > 0$ such that $\|x(t; t_0, \psi) - \bar{x}(t; t_0, \varphi)\|_\infty \leq \varepsilon$ for all $t \geq t_0$ and for any ψ such that $\|\psi - \varphi\|_X \leq \delta$. $\bar{x}$ is called uniformly stable when δ is independent of t_0 and unstable when it is not stable.

Definition 2.4 (*asymptotically stable solution*) The solution $\bar{x}(\cdot; t_0, \varphi)$ of (2.9) is called asymptotically stable if it is stable and, in addition, there exists $\delta = \delta(t_0) > 0$ such that $\|x(t; t_0, \psi) - \bar{x}(t; t_0, \varphi)\|_\infty \to 0$ as $t \to +\infty$ for any ψ such that $\|\psi - \varphi\|_X \leq \delta$. $\bar{x}$ is called uniformly asymptotically stable when it is uniformly stable and there is $r > 0$ such that for every $\gamma > 0$ there is $t_f(\gamma) > t_0$ such that $\|x(t; t_0, \psi) - \bar{x}(t; t_0, \varphi)\|_\infty \leq \gamma$ for any $t_0 \in \mathbb{R}$, $t \geq t_f(\gamma)$ and ψ such that $\|\psi - \varphi\|_X \leq r$.

Uniform asymptotic stability means that $\|x(t; t_0, \psi) - \bar{x}(t; t_0, \varphi)\|_\infty \to 0$ uniformly w.r.t ψ whenever $\|\psi - \varphi\|_X \leq r$ and, in general, is stronger than stability. For some DDEs they coincide, e.g., for autonomous and periodic ones [91, Chap. 5, Lemma 1.1]. The same holds also for asymptotic stability. We recall a further definition: $\bar{x}(t; t_0, \varphi)$ is called locally *exponentially* stable if there are positive constants α_1, α_2 and β such that $\|x(t; t_0, \psi) - \bar{x}(t; t_0, \varphi)\|_\infty \leq \alpha_1 e^{-\alpha_2(t-t_0)}$ for all $t \geq t_0$ and for any ψ such that $\|\psi - \varphi\|_X \leq \beta$. In any case, all definitions reflect the local nature of stability: if we slightly perturb the initial function, then the perturbed solution stays in the neighborhood of $\bar{x}$ or, for asymptotic stability, returns to it.

2.3 Stability of Solutions

Now, assume that F is continuously differentiable w.r.t. its second argument and denote by DF its Fréchet derivative [8]. To examine the stability of a specific solution $\bar{x}(\cdot; t_0, \varphi)$ of (2.9), we apply the *Principle of Linearized Stability*: the study of the stability of a solution $\bar{x}$ is reduced to the study of the system linearized at $\bar{x}$, i.e.,

$$x'(t) = DF(t, \bar{x}_t)x_t, \; t \geq t_0. \tag{2.11}$$

System (2.11), also called *variational* in the literature, can be viewed as a first-order approximation of (2.1). In the linearization approach, the starting point is the study of the behavior of the zero solution of linear DDEs, which is complemented with suitable theorems asserting that the local behavior of the solutions close to $\bar{x}$ is determined, to the first order, by the behavior of the solutions of (2.11). It is important to underline that, in some critical situations, one cannot conclude anything without investigating higher order terms [70, Chap. IX]. According to this principle, the analysis of linear DDEs is essential: we present the theoretical aspects of the stability of linear autonomous DDEs in Chap. 3 and of linear periodic DDEs in Chap. 4, laying the foundation for the numerical methods of Part II.

Chapter 3
Stability of Linear Autonomous Equations

We focus our attention on linear autonomous DDEs and on the analysis of the stability properties of the zero solution. To this aim we consider the semigroup approach. By introducing the infinitesimal generator associated to the family of solution operators, we describe the dynamics of the state in the infinite dimensional state space by an abstract ODE (Sect. 3.1). Similarly to the finite dimensional case, the spectral properties of the infinitesimal generator give the conditions for the stability of the zero solution. A characteristic equation is also derived, whose roots are the eigenvalues of the infinitesimal generator (Sect. 3.2). The stability can be carried out also by analyzing the spectrum of the solution operator. The latter is generally not known explicitly, hence it needs to be approximated numerically in order to estimate its spectrum. Therefore, from a theoretical point of view, the possibility to analyze the spectrum of the infinitesimal generator, which is known, as well as the presence of a characteristic equation, represent the main advantages of the theory of semigroups applied to linear autonomous DDEs. Moreover, it indicates an alternative path to follow for the stability analysis by numerical methods, which are in any case necessary due to the complexity of the problem (mainly the infinite dimension). In Sect. 3.3 we study the local stability of equilibria of nonlinear autonomous DDEs by following the Principle of Linearized Stability and give the relevant results, for which the understanding of the linear autonomous case is crucial. Throughout the chapter, we present some basic results and we prove only some theorems specific for DDEs, leaving all the details of a general semigroup theory to [74] and referring to [15, 70, 91, 93] for the case of DDEs. Interested readers find applications of the semigroup approach to other types of evolution equations in [62, 74, 156, 204].

Consider the linear autonomous DDE (2.7), i.e.,

$$x'(t) = Lx_t, \quad t \in \mathbb{R}, \tag{3.1}$$

where $L : X \to \mathbb{R}^d$ is a linear and bounded functional. Theorem 2.2 ensures that

$$\begin{cases} x'(t) = Lx_t, \ t \geq 0, \\ x_0 = \varphi \end{cases} \tag{3.2}$$

has a unique global solution, which admits the following representation:

$$x(t;\varphi) = \begin{cases} \varphi(0) + \int_0^t Lx_s ds & \text{if } t \geq 0, \\ \varphi(t) & \text{if } t \in [-\tau, 0]. \end{cases} \quad (3.3)$$

3.1 The Solution Operator Semigroup and the Infinitesimal Generator

Having the well-posedness of (3.2), we can face the study of the qualitative properties of the solution. To this aim, the theory of one-parameter semigroups represents a powerful mathematical tool. The focus is on the infinitesimal generator, which allows to introduce an abstract ODE describing the dynamics of the state in the infinite dimensional state space X. We first introduce the necessary basic concepts in general for a family $\{T(t)\}_{t \geq 0}$ of linear and bounded operators $T(t) : Y \to Y$ on a Banach space $(Y, \|\cdot\|_Y)$. Then we go through DDEs, laying the theoretical basis for the construction of the numerical approach presented in Chap. 5.

Definition 3.1 (*strongly continuous semigroup*) A family $\{T(t)\}_{t \geq 0}$ of linear and bounded operators $T(t) : Y \to Y$ on a Banach space Y is called a strongly continuous semigroup (or C_0-semigroup) whenever it satisfies

- the semigroup properties: $T(0) = I_Y$ and $T(t+s) = T(t)T(s)$ for all $t, s \geq 0$;
- the strong continuity property: for any $\varphi \in Y$, $\|T(t)\varphi - \varphi\|_Y \to 0$ as $t \downarrow 0$.

Definition 3.2 (*infinitesimal generator*) Let $\{T(t)\}_{t \geq 0}$ be a C_0-semigroup of linear and bounded operators on Y. The operator $\mathscr{A} : \mathscr{D}(\mathscr{A}) \subseteq Y \to Y$ defined as

$$\begin{cases} \mathscr{D}(\mathscr{A}) = \left\{ \varphi \in Y : \lim_{h \downarrow 0} \dfrac{T(h)\varphi - \varphi}{h} \text{ exists in } Y \right\} \\ \mathscr{A}\varphi = \lim_{h \downarrow 0} \dfrac{T(h)\varphi - \varphi}{h} \end{cases} \quad (3.4)$$

is called the infinitesimal generator of $\{T(t)\}_{t \geq 0}$.

The infinitesimal generator represents the (right-hand) derivative of $T(t)$ in $t = 0$. It is a linear, closed, densely defined and, in general, unbounded operator, uniquely defined by the C_0-semigroup. The following fundamental result links semigroups and dynamical systems from abstract ODEs through the infinitesimal generator.

Theorem 3.1 *Let $\{T(t)\}_{t \geq 0}$ be a C_0-semigroup of linear and bounded operators on Y with infinitesimal generator $\mathscr{A}$. For any $\varphi \in \mathscr{D}(\mathscr{A})$, the function $u : t \mapsto u(t) := T(t)\varphi$, $t \geq 0$, is the unique (classic) solution of the abstract Cauchy problem on Y*

3.1 The Solution Operator Semigroup and the Infinitesimal Generator

$$\begin{cases} u'(t) = \mathscr{A} u(t), \ t \geq 0, \\ u(0) = \varphi, \end{cases} \tag{3.5}$$

i.e., $u(t)$ is continuously differentiable and $u(t) \in \mathscr{D}(\mathscr{A})$ for all $t \geq 0$ and (3.5) holds.

We remark that the classic solution of (3.5) requires $\varphi \in \mathscr{D}(\mathscr{A})$. It is possible to define a mild solution of (3.5) for all $\varphi \in Y$ [74, Chap. II, Definition 6.3]).

Now, focusing on DDEs, bearing in mind that in (3.2) the initial state φ belongs to the Banach space $(X, \|\cdot\|_X)$ and that the state at time t is the function $x_t \in X$ in (2.2), we first give the following definition to trace the time evolution of the state in the state space and then we apply the above theory of one-parameter semigroups.

Definition 3.3 (*solution operator*) The operator $T(t) : X \to X$ associating to the initial function $\varphi \in X$ the state x_t at time $t \geq 0$ by (3.2), i.e.,

$$T(t)\varphi = x_t(\cdot; \varphi), \tag{3.6}$$

is called the solution operator.

Proposition 3.1 *The family $\{T(t)\}_{t \geq 0}$ of solution operators (3.6) defines a C_0-semigroup of linear and bounded operators on X.*

Proof The linearity of $T(t)$ in (3.6) easily follows from the linearity of (3.1). To prove boundedness, we express $T(t)$ for all $t \geq 0$ and $\theta \in [-\tau, 0]$ through (3.3) as

$$(T(t)\varphi)(\theta) = \begin{cases} \varphi(0) + \displaystyle\int_0^{t+\theta} LT(s)\varphi \, ds & \text{if } t + \theta \geq 0, \\ \varphi(t+\theta) & \text{if } t + \theta \leq 0. \end{cases} \tag{3.7}$$

The boundedness of L implies $\|T(t)\varphi\|_X \leq \|\varphi\|_X + \int_0^t \|L\| \|T(s)\varphi\|_X ds$, $t \geq 0$, and, by Gronwall's inequality [21, 88], we get

$$\|T(t)\varphi\|_X \leq \|\varphi\|_X e^{\|L\|t}, \ t \geq 0. \tag{3.8}$$

Consequently, the bound $\|T(t)\| \leq e^{\|L\|t}$, $t \geq 0$, holds. The semigroup properties in Definition 3.1 easily follow from (3.6) and from the uniqueness of the solution of (3.2). As for the strong continuity, the solution $x(t; \varphi)$ of (3.2) is continuous for $t \geq -\tau$ and, therefore, it is uniformly continuous on bounded intervals $[-\tau, t_e]$ for any $t_e > 0$. Hence, for any $\varepsilon > 0$, there exists $\delta > 0$ such that $\|x(t_1; \varphi) - x(t_2; \varphi)\|_\infty < \varepsilon$ if $t_1, t_2 \in [-\tau, t_e]$ are such that $|t_1 - t_2| < \delta$. Consequently, for $0 \leq t < \delta$, we obtain $\|x(t+\theta; \varphi) - x(\theta; \varphi)\|_\infty = \|x(t+\theta; \varphi) - \varphi(\theta)\|_\infty < \varepsilon$ for all $\theta \in [-\tau, 0]$, showing the strong continuity and thus completing the proof. □

Strong continuity implies the continuity of the map $t \to T(t)\varphi$ for any $\varphi \in X$.

Hereafter, we refer to the C_0-semigroup of solution operators (3.6) simply as the *SO-semigroup*. We now characterize the infinitesimal generator associated to the latter: it is a derivative operator subject to a constraint imposed by (3.1).

Proposition 3.2 *The infinitesimal generator of the SO-semigroup is the linear unbounded operator* $\mathscr{A} : \mathscr{D}(\mathscr{A}) \subseteq X \to X$ *given by*

$$\begin{cases} \mathscr{D}(\mathscr{A}) = \{\varphi \in X \,:\, \varphi' \in X,\ \varphi'(0) = L\varphi\}, \\ \mathscr{A}\varphi = \varphi'. \end{cases} \tag{3.9}$$

Proof Let $\varphi \in \mathscr{D}(\mathscr{A})$ and denote $\psi = \mathscr{A}\varphi \in X$. According to (3.4) we have

$$\lim_{h \downarrow 0} \left\| \frac{T(h)\varphi - \varphi}{h} - \psi \right\|_X = \lim_{h \downarrow 0} \max_{\theta \in [-\tau, 0]} \left\| \frac{(T(h)\varphi)(\theta) - \varphi(\theta)}{h} - \psi(\theta) \right\|_\infty = 0.$$

Take $\theta \in [-\tau, 0)$. Since $\theta + h < 0$ for $h \downarrow 0$, we have

$$\psi(\theta) = \lim_{h \downarrow 0} \frac{T(h)\varphi(\theta) - \varphi(\theta)}{h} = \lim_{h \downarrow 0} \frac{\varphi(\theta + h) - \varphi(\theta)}{h}.$$

For $\theta = 0$ we get from (3.7)

$$\psi(0) = \lim_{h \downarrow 0} \frac{T(h)\varphi(0) - \varphi(0)}{h} = \lim_{h \downarrow 0} \frac{\int_0^h LT(s)\varphi\,ds}{h} = L\varphi.$$

Therefore, φ is right-differentiable on $[-\tau, 0)$ with right-hand derivative ψ and, moreover, $\psi(0) = L\varphi$. From

$$\left\| \frac{\varphi(\theta - h) - \varphi(\theta)}{-h} - \psi(\theta) \right\|_\infty \leq \left\| \frac{\varphi(s + h) - \varphi(s)}{h} - \psi(s) \right\|_\infty + \|\psi(s) - \psi(s + h)\|_\infty,$$

where $s = \theta - h$ and both the addends to the right converge uniformly in s by the continuity of ψ, we conclude that φ is differentiable and $\varphi'(0) = L\varphi$. Conversely, suppose that $\varphi' \in X$ and $\varphi'(0) = L\varphi$. By defining $\psi = \varphi'$ we have that

$$\left\| \frac{\varphi(\theta + h) - \varphi(\theta)}{h} - \psi(\theta) \right\|_\infty = \left\| \frac{1}{h} \int_0^h (\psi(\theta + s) - \psi(\theta))\,ds \right\|_\infty$$

converges uniformly to zero as $h \downarrow 0$ for $\theta \in [-\tau, 0]$. This concludes the proof. □

By Theorem 3.1, the qualitative and, in particular, the asymptotic behavior of the solutions of (3.1) are described by the dynamics of the SO-semigroup through the crucial reformulation of (3.1) as the linear abstract ODE on X

$$u'(t) = \mathscr{A}u(t), \ t \geq 0, \tag{3.10}$$

for $\mathscr{A}$ the infinitesimal generator (3.9). Combining the results above, we conclude that $T(t)\varphi = x_t(\cdot; \varphi)$ in (3.6) solves (3.5) for $\mathscr{A}$ in (3.9). To study the dynamics of the SO-semigroup, we investigate in the following section the spectrum of the solution operator and its relation with the spectrum of its infinitesimal generator.

Remark 3.1 The spectral analysis requires to work on Banach spaces on $\mathbb{C}$. Hereafter (and in Part II) we implicitly assume that X and all the operators involved have been complexified [70]. In Part III, about tests and applications, we go back to $\mathbb{R}$.

3.2 Spectral Properties and the Characteristic Equation

The abstract ODE (3.10) is infinite dimensional but, similarly to the finite dimensional case, it suggests that the asymptotic properties of the solutions of the linear autonomous DDE (3.1) and the stability of the zero solution depend on the spectrum of $\mathscr{A}$. In general, the spectrum of an infinite dimensional operator exhibits a more rich structure than that of a matrix. Our aim here is to describe the spectrum of both $T(t)$ and $\mathscr{A}$ defined in (3.6) and (3.9), respectively, and their relation. In the particular case we are dealing with, the eventual compactness of the SO-semigroup affects the spectrum of both operators (which contain only eigenvalues) and, therefore, the investigation of the asymptotic behavior of the solutions and the stability of the zero solution. Moreover, it allows to determine the behavior of the SO-semigroup by the location in $\mathbb{C}$ of the spectrum of its infinitesimal generator. It is an important result in applications, since the stability can be analyzed without solving the equation.

In the same spirit of the previous section, we first give the basic definitions and results for linear operators and C_0-semigroups on a general Banach space Y. Eventually, we specialize to the SO-semigroup for DDEs on the state space X.

Definition 3.4 (*resolvent and spectrum*) Let $\mathscr{A} : \mathscr{D}(\mathscr{A}) \subseteq Y \to Y$ be a linear (closed or bounded) operator. The resolvent set of $\mathscr{A}$ is $\rho(\mathscr{A}) = \{\lambda \in \mathbb{C} : \lambda I_Y - \mathscr{A} \text{ is bijective}\}$. The spectrum of $\mathscr{A}$ is the complementary set $\sigma(\mathscr{A}) = \mathbb{C} \setminus \rho(\mathscr{A})$.

The resolvent set $\rho(\mathscr{A})$ is open in $\mathbb{C}$ and, hence, the spectrum $\sigma(\mathscr{A})$ is closed in $\mathbb{C}$.

Definition 3.5 (*point spectrum*) The point spectrum $\sigma_P(\mathscr{A})$ of $\mathscr{A}$ is the set of $\lambda \in \mathbb{C}$ such that $\lambda I_Y - \mathscr{A}$ is not injective, i.e., $\mathscr{A}\varphi = \lambda\varphi$ for some $\varphi \neq 0$. We call λ an eigenvalue and φ the corresponding eigenfunction.

Let λ be an eigenvalue of $\mathscr{A}$. The null space $\mathscr{N}(\lambda I_Y - \mathscr{A})$ is called the *eigenspace* of λ and its dimension $g(\lambda)$ is called the *geometric multiplicity*. The smallest closed linear subspace $\mathscr{E}(\lambda)$ that contains all $\mathscr{N}(\lambda I_Y - \mathscr{A})^k$ for $k \geq 1$ is called the *generalized eigenspace* of λ and its dimension $\nu(\lambda)$ is called the *algebraic multiplicity*. If λ is an isolated point of $\sigma_P(\mathscr{A})$ and $\nu(\lambda) < +\infty$, then λ is called an eigenvalue of *finite type* and *simple* if $\nu(\lambda) = 1$. For an isolated λ, the smallest number $\ell(\lambda)$ such that

$\mathscr{E}(\lambda) = \mathscr{N}(\lambda I_Y - \mathscr{A})^{\ell(\lambda)}$ is called the *ascent*. In the finite dimensional case, the spectrum contains only eigenvalues, whereas for infinite dimensional operators this is not true in general. But the spectrum of a *compact* operator, which maps bounded sets into relatively compact sets, has a simple structure [124, Theorem 3.11]. In fact, it is a countable set, which can only accumulate at zero. Moreover, any nonzero element in the spectrum is an isolated eigenvalue with finite algebraic multiplicity and, therefore, it has properties quite similar to the eigenvalues of a finite dimensional operator. The following definition emphasizes this fundamental property.

Definition 3.6 (*eventually compact semigroup*) A C_0-semigroup $\{T(t)\}_{t\geq 0}$ of linear and bounded operators on Y is called eventually compact if there exists $\bar{t} > 0$ such that $T(\bar{t})$ is compact.

In contrast to the finite dimensional case, in the infinite dimensional case (3.5) one should distinguish between the case where all solutions decay exponentially and the case where only the classic ones do. The following notion characterizes the long-time behavior of all the solutions.

Definition 3.7 (*growth bound*) Let $\{T(t)\}_{t\geq 0}$ be a C_0-semigroup of linear and bounded operators on Y. The real number

$$\omega_0 := \inf\{\omega : \text{ there exists } M_\omega > 0 \text{ such that } \|T(t)\| \leq M_\omega e^{\omega t} \text{ for all } t \geq 0\}$$

is called the growth bound of $\{T(t)\}_{t\geq 0}$.

We are interested to learn about the growth bound ω_0 from the spectrum of the infinitesimal generator. The fundamental relation between the spectrum of each operator $T(t)$ of an eventually compact C_0-semigroup $\{T(t)\}_{t\geq 0}$ and the spectrum of its generator is stated in the following Spectral Mapping Theorem.

Theorem 3.2 *Let $\{T(t)\}_{t\geq 0}$ be a C_0-semigroup of linear and bounded operators on Y with infinitesimal generator $\mathscr{A}$. If $\{T(t)\}_{t\geq 0}$ is eventually compact then*

$$\sigma(T(t)) \setminus \{0\} = e^{t\sigma(\mathscr{A})}, \quad t \geq 0. \tag{3.11}$$

Having in mind (3.11) we introduce the following definition.

Definition 3.8 (*spectral abscissa*) Let $\mathscr{A} : \mathscr{D}(\mathscr{A}) \subseteq Y \to Y$ be a linear (unbounded) closed operator. The constant $s(\mathscr{A}) := \sup\{\operatorname{Re}(\lambda) : \lambda \in \sigma(\mathscr{A})\}$ is called the spectral abscissa of $\mathscr{A}$.

We remark that $s(\mathscr{A})$ is also called spectral bound by some authors (see, e.g., [74, Definition 2.1]). Here we adopt the definition in [70]. In general, $s(\mathscr{A}) \leq \omega_0$. For eventually compact C_0-semigroups, equality holds.

Theorem 3.3 *Let $\{T(t)\}_{t\geq 0}$ be an eventually compact C_0-semigroup of linear and bounded operators on Y with infinitesimal generator $\mathscr{A}$. Then $\omega_0 = s(\mathscr{A})$.*

As anticipated, we now focus on the SO-semigroup for linear autonomous DDEs, first showing its compactness properties.

3.2 Spectral Properties and the Characteristic Equation

Proposition 3.3 *The SO-semigroup is eventually compact. In particular, $T(t)$ in (3.6) is compact for all $t \geq \tau$. Moreover,*

$$\sigma(T(t)) \subseteq \sigma_p(T(t)) \cup \{0\}, \quad t \geq 0. \tag{3.12}$$

Proof We prove that $T(t)$ maps bounded sets into relatively compact sets for $t \geq \tau$. Let $B = \{\varphi \in X : \|\varphi\|_X \leq \beta\}$, $\beta > 0$. For any $\psi \in T(t)B$, $t \geq \tau$, (3.8) implies $\|\psi\|_X \leq e^{\|L\|t}\beta$, whereas (3.1) implies $\|\psi'\|_X \leq \|L\|e^{\|L\|t}\beta$. Since any ψ is uniformly bounded with uniformly bounded derivatives, the compactness of $T(t)B$ follows by the Ascoli–Arzelà Theorem [124, Theorem 1.18]. For $t = 0$ the relation (3.12) holds. For any $t > 0$ there exists $q \in \mathbb{N}$ such that $qt \geq \tau$ and, therefore, the operator $T(qt) = T(t)^q$ is compact and $\sigma(T(t)^q) \subseteq \sigma_p(T(t)^q) \cup \{0\}$. Since $T(t)$ is bounded, $\sigma(T(t)^q) = \sigma(T(t))^q$ [70, Appendix II, Exercise 4.8], proving (3.12). □

By combining the latter proposition with all the previous general results, we obtain the following theorem, which extends to linear autonomous DDEs the analogous one valid for linear autonomous ODEs.

Theorem 3.4 *Let $\{T(t)\}_{t\geq 0}$ be the SO-semigroup with infinitesimal generator $\mathscr{A}$ as given in (3.9). The following statements are equivalent:*

- *the zero solution of (3.1) is asymptotically exponentially stable;*
- $\sigma(T(t)) \subseteq \{\mu \in \mathbb{C} : |\mu| < 1\}$;
- $s(\mathscr{A}) < 0$.

The asymptotic behavior of the SO-semigroup, represented by the growth bound ω_0, can be inferred from the spectral properties of its generator, represented by the spectral abscissa $s(\mathscr{A})$. This characterization is of particular importance since, unlike the infinitesimal generator, an explicit form of the solution operator is generally not known as already remarked. Hence, we focus our attention on the generator of the SO-semigroup proving that, similarly to the case of ODEs, its spectrum (containing only isolated eigenvalues) coincides with the roots of a characteristic equation.

Proposition 3.4 *Let $\mathscr{A}$ given in (3.9) be the infinitesimal generator of the SO-semigroup. Then $\sigma(\mathscr{A})$ contains only eigenvalues and $\lambda \in \sigma_P(\mathscr{A})$ if and only if λ satisfies the characteristic equation*

$$\det(\Delta(\lambda)) = 0, \tag{3.13}$$

where

$$\Delta(\lambda) := \lambda I_d - L(e^{\lambda \cdot}) \tag{3.14}$$

and $L(e^{\lambda \cdot})u = L(e^{\lambda \cdot}u)$, $u \in \mathbb{C}^d$. The eigenvalues are of finite-type, with real part bounded above and any vertical strip of $\mathbb{C}$ contains only a finite number of them.

Proof We first characterize $\sigma_P(\mathscr{A})$. From (3.9) we have that an eigenfunction φ associated to $\lambda \in \sigma_P(\mathscr{A})$ is represented by

$$\varphi(\theta) = \varphi(0)e^{\lambda\theta}, \tag{3.15}$$

with $\varphi(0) \in \mathbb{C}^d \setminus \{0\}$. Moreover, φ belongs to the domain $\mathscr{D}(\mathscr{A})$ if and only if it satisfies the domain condition $\varphi'(0) = L\varphi$, which becomes

$$\lambda\varphi(0) = L(e^{\lambda\cdot})\varphi(0). \tag{3.16}$$

Therefore, by defining $\Delta(\lambda)$ as in (3.14), we obtain that $\sigma_P(\mathscr{A}) = \{\lambda \in \mathbb{C} : \det(\Delta(\lambda)) = 0\}$. Since $\sigma(\mathscr{A})$ is the complement of $\rho(\mathscr{A})$, we now show that the resolvent set $\rho(\mathscr{A})$ contains any λ not satisfying (3.13). $\lambda \in \rho(\mathscr{A})$ if and only if

$$(\mathscr{A} - \lambda I)\varphi = \psi \tag{3.17}$$

has a solution $\varphi \in \mathscr{D}(\mathscr{A})$ for every ψ in a dense set in X, which depends continuously on such ψ. We solve (3.17) for φ given $\psi \in X$. Proceeding as above, from (3.9) we have that a solution φ of (3.17) can be represented as

$$\varphi(\theta) = e^{\lambda\theta}\varphi(0) + \int_0^\theta e^{\lambda(\theta-s)}\psi(s)ds, \ \theta \in [-\tau, 0], \tag{3.18}$$

where $\varphi(0)$ satisfies $\Delta(\lambda)\varphi(0) = \psi(0) - L\left(\int_0^\cdot e^{\lambda(\cdot-s)}\psi(s)ds\right)$. From the latter and from (3.18) we have that (3.17) has a solution, which is continuous w.r.t. ψ, if and only if $\det(\Delta(\lambda)) \neq 0$. Therefore, $\rho(\mathscr{A}) = \{\lambda \in \mathbb{C} : \det(\Delta(\lambda)) \neq 0\}$. The last part of the theorem can be proved by using classic results from complex analysis and the theory of linear operators, see [91, p.169] and [70, Chap. IV, Exercise 2.8]. □

Let us note that the above results, but for the characteristic equation, can be proved by using general facts holding for eventually compact C_0-semigroups.

In the latter proof, we have seen that an eigenfunction φ associated to $\lambda \in \sigma_P(\mathscr{A})$ is given by (3.15), with $\varphi(0)$ a nonzero solution of (3.16). From $\frac{d}{dt}T(t)\varphi = T(t)\mathscr{A}\varphi = \lambda T(t)\varphi$ [74, Chap. II, Lemma 1.3], we get $T(t)\varphi = \varphi e^{\lambda t}$ for $t \geq 0$. Hence, φ is also the eigenfunction associated to the eigenvalue $\mu = e^{\lambda t}$ of the solution operator $T(t)$ (see, indeed, Theorem 3.2). Also, the solution of (3.2) with initial function φ is $x_t(\theta; \varphi) = \varphi(0)e^{\lambda(t+\theta)}, t \geq 0, \theta \in [-\tau, 0]$. In the following proposition, we find the expression of the solution of (3.2) with initial function in the generalized eigenspace $\mathscr{E}(\lambda)$ and we prove that on it (3.1) behaves essentially like an ODE. The result can be generalized to a finite number of eigenvalues.

Proposition 3.5 *Let λ be an eigenvalue with multiplicity $\nu(\lambda)$ of $\mathscr{A}$ given in (3.9) and let $\Phi = (\varphi_1, \ldots, \varphi_{\nu(\lambda)})$ be a basis for the generalized eigenspace $\mathscr{E}(\lambda)$. The solutions of (3.2) with initial function $\varphi = \Phi b \in \mathscr{E}(\lambda)$, with b a $\nu(\lambda)$-dimensional vector, are of the form $x_t(\theta; \varphi) = \Phi e^{C(t+\theta)}b, t \geq 0, \theta \in [-\tau, 0]$, where C is a $\nu(\lambda)$-dimensional matrix with λ as the only eigenvalue.*

3.2 Spectral Properties and the Characteristic Equation

Proof For any eigenvalue λ of $\mathscr{A}$, $\mathscr{E}(\lambda)$ is finite dimensional with dimension $\nu(\lambda)$. Let $\varphi_i \in X$, $i = 1, \ldots, \nu(\lambda)$ be a basis for $\mathscr{E}(\lambda)$. Define the row vector $\Phi = (\varphi_1, \ldots, \varphi_{\nu(\lambda)}) \in X^{\nu(\lambda)}$. Since $\mathscr{A}\mathscr{E}(\lambda) \subseteq \mathscr{E}(\lambda)$, there exists a matrix C of dimension $\nu(\lambda)$ such that $\mathscr{A}\Phi = \Phi C$, where $\mathscr{A}\Phi = (\mathscr{A}\varphi_1, \ldots, \mathscr{A}\varphi_{\nu(\lambda)})$. The only eigenvalue of C is λ. By (3.9) we get $\Phi(\theta) = \Phi(0)e^{C\theta}$, $\theta \in [-\tau, 0]$. Since $\varphi_i \in \mathscr{D}(\mathscr{A})$, $i = 1, \ldots, \nu(\lambda)$, we have that $T(t)\varphi_i$ is differentiable and $\frac{d}{dt}(T(t)\varphi_i) = T(t)\mathscr{A}\varphi_i$. Hence, $\frac{d}{dt}T(t)\Phi = T(t)\mathscr{A}\Phi = T(t)\Phi C$, which implies $T(t)\Phi = \Phi e^{Ct}$ for $t \geq 0$. For $\varphi = \Phi b \in \mathscr{E}(\lambda)$, a combination of the previous gives $x_t(\theta; \varphi) = T(t)\Phi(\theta)b = \Phi(\theta)e^{Ct}b = \Phi(0)e^{C(t+\theta)}b$, $\theta \in [-\tau, 0]$, and the result follows. $\square$

Eventually, concerning applications of interest, we consider the prototype model of linear autonomous DDEs (2.8), whose characteristic equation reads

$$\det\left(\lambda I_d - A - \sum_{k=1}^{p} B_k e^{-\lambda \tau_k} - \sum_{k=1}^{p} \int_{-\tau_k}^{-\tau_{k-1}} C_k(\theta)e^{\lambda\theta}d\theta\right) = 0. \quad (3.19)$$

In general, there are not necessary and sufficient conditions for all the roots of (3.19) to be in the left half-plane. Therefore, the location of the roots and the construction of stability charts is a difficult task, often requiring efficient numerical approaches. Only for some simple classes of linear autonomous DDEs, the study can be developed by analytical methods, e.g., the Hayes equation (1.4) [33].

3.3 Linearization and Equilibria

Consider the nonlinear autonomous DDE (2.3) and assume that F is continuous with bounded Fréchet derivative. Theorem 2.2 ensures that (2.10) has a unique solution on $[-\tau, +\infty)$. In applications, the solutions which are independent of time are particularly important since they represent a behavior which persists in time.

Definition 3.9 (*equilibrium*) An equilibrium $\bar{x} \in X$ for (2.3) is a constant mapping with value $\bar{x}$ which satisfies $F(\bar{x}) = 0$.

Clearly, an equilibrium is a solution of (2.3). We remark that, in general, numerical methods may be necessary to find the zeros of F.

From a dynamical system point of view, it is important to analyze whether or not the solutions nearby an equilibrium remain nearby, get closer or go far away. Definition 2.3 apply straightforwardly. There can be more than one equilibrium, with different stability properties. According to the Principle of Linearized Stability, we can reduce the investigation of the local stability properties of an equilibrium $\bar{x}$ to the stability analysis of the zero solution of the linearized system

$$x'(t) = DF(\bar{x})x_t, \quad t \geq 0, \quad (3.20)$$

where $DF(\bar{x})$ is the Fréchet derivative of F at $\bar{x}$. To complete the analysis, we need a result linking the stability properties of $\bar{x}$ to that of the zero solution of (3.20). In other words, the Principle of Linearized Stability is accomplished in the following.

Theorem 3.5 *Let $\bar{x}$ be an equilibrium of* (2.3), $\{T(t)\}_{t\geq 0}$ *the SO-semigroup and $\mathscr{A}$ the infinitesimal generator associated to* (3.20). *Then $\bar{x}$ is asymptotically exponentially stable if the zero solution of* (3.20) *is asymptotically exponentially stable, that is $s(\mathscr{A}) < 0$ or, equivalently, $\sigma(T(t)) \subseteq \{\mu \in \mathbb{C} : |\mu| < 1\}$, $t > 0$. $\bar{x}$ is unstable if the zero solution of* (3.20) *is unstable, i.e., there is at least one $\lambda \in \sigma(\mathscr{A})$ such that $\mathrm{Re}(\lambda) > 0$ or, equivalently, there is at least one $\mu \in \sigma(T(t))$ such that $|\mu| > 1$.*

Note that the case when some eigenvalues of $\mathscr{A}$ have zero real part or, equivalently, some eigenvalues of $T(t)$ have unitary modulus, is not covered by Theorem 3.5. In this case, the investigation of stability requires a deeper analysis [70, Chap. IX].

Theorem 3.5 indicates two alternatives: to determine the position in $\mathbb{C}$ of the infinitely many eigenvalues of either the solution operator or the infinitesimal generator of (3.20). But, as already pointed out, in general the solution operator has not an explicit form, whereas the infinitesimal generator is known and, moreover, its eigenvalues solve the characteristic equation (3.13) as proved in Proposition 3.4. From a numerical point of view, both the alternatives have been followed (see Part II and the references therein). We conclude with an example illustrating the linearized approach for the stability analysis of equilibria.

Example 3.1 The Mackey-Glass equation is the nonlinear DDE

$$x'(t) = \beta \frac{x(t-\tau)}{1+x(t-\tau)^n} - \gamma x(t), \ t \geq 0, \quad (3.21)$$

with β, γ, τ and n positive real numbers. It has been proposed in [138] to study the dynamics of physiological systems, such as the density of mature circulating white blood cells, where τ is the delay between the cells production and maturation and release into the bloodstream. It is also a celebrated example of chaos for DDEs. We now focus on the equilibria, i.e., the solutions $\bar{x}$ of

$$x\left(\frac{\beta}{1+x^n} - \gamma\right) = 0, \quad (3.22)$$

which are $\bar{x}_0 = 0$, existing for all the values of the parameters, and $\bar{x}_1 = \sqrt[n]{r-1}$ for $r := \beta/\gamma$, existing positive only when $r > 1$ and coinciding with $\bar{x}_0$ when $r = 1$. By rewriting (3.21) as $x'(t) = F(x(t), x(t-\tau))$, the variational equation is obtained as

$$x'(t) = \frac{\partial F}{\partial x(t)}(\bar{x},\bar{x})x(t) + \frac{\partial F}{\partial x(t-\tau)}(\bar{x},\bar{x})x(t-\tau) \quad (3.23)$$

for $\bar{x}$ either $\bar{x}_0$ or $\bar{x}_1$. At $\bar{x}_0$, (3.23) becomes $x'(t) = -\gamma x(t) + \beta x(t-\tau)$ and by using the stability results summarized in Sect. 1.2 (Fig. 1.1), we conclude that $\bar{x}_0$ is

3.3 Linearization and Equilibria

asymptotically stable for $0 < r < 1$, whereas $r = 1$ is a steady-state bifurcation point. When $r > 1$, $\bar{x}_0$ is unstable and, in fact, (3.22) has the positive solution $\bar{x}_1$. The linearized equation at $\bar{x}_1$ is

$$x'(t) = -\gamma x(t) + \gamma \left(\frac{r - n(r-1)}{r} \right) x(t - \tau). \tag{3.24}$$

When $n = 1$ we easily conclude that $\bar{x}_1$ is asymptotically stable. Instead, if $n > 1$, we have $0 < \frac{r-n(r-1)}{r} < 1$ for $1 < r < \frac{n}{n-1}$ and by using again the results in Sect. 1.2, we conclude that $\bar{x}_1$ is asymptotically stable for $1 < r \leq \frac{n}{n-1}$. For $r > \frac{n}{n-1}$ the coefficient of the delayed term becomes negative and the asymptotic stability is preserved until the coefficients of (3.24) are in the stability domain (Fig. 1.1). Also the parameter τ plays an important role [83].

Chapter 4
Stability of Linear Periodic Equations

The central subject of this chapter is the stability analysis of the zero solution of linear periodic DDEs. A theory similar to the Floquet one for linear periodic ODEs has been developed. In the sequel, we emphasize both the similarities and the essential differences w.r.t. the linear autonomous case. Indeed, an analogous of the infinitesimal generator is not well-defined and the monodromy operator with its characteristic multipliers plays the crucial role. As the linearized stability theory has been successfully applied to the equilibria of autonomous DDEs in Chap. 3, by combining the linearized approach with the Floquet theory we also relate the stability of a nonconstant periodic solution of a nonlinear autonomous DDE to the position w.r.t. the unit circle in $\mathbb{C}$ of the characteristic multipliers of the linearized system.

We refer to [70, 91] for a complete treatment of the Floquet theory for linear periodic DDEs. Here we present the basic results useful for the numerical approach developed in Chap. 6. Let us note that the latter is valid also in the more general case of linear nonautonomous (not necessarily periodic) DDEs (2.4). Hence, in Sect. 4.1, we start from the notions of evolution operators and families: they generalize those of solution operators and one-parameter semigroups to two parameters (see [62, Chap. 3] for a general reference). Then, a particular instance of evolution operator is the monodromy operator associated to linear periodic DDEs (2.6), i.e.,

$$x'(t) = L(t)x_t, \quad t \in \mathbb{R}, \tag{4.1}$$

where $L(t)$, $t \in \mathbb{R}$, is a family of linear, bounded, and periodic functionals: there is an $\omega > 0$ such that $L(t + \omega) = L(t)$ for all $t \in \mathbb{R}$. In both cases (periodic or not), Theorem 2.2 ensures, for any s (in $\mathbb{R}$ or in $\mathbb{I}$, respectively) and $\varphi \in X$, the existence of a unique solution $x(\cdot; s, \varphi)$ on $[s - \tau, +\infty)$ of

$$\begin{cases} x'(t) = L(t)x_t, \quad t \geq s, \\ x_s = \varphi. \end{cases} \tag{4.2}$$

4.1 The Evolution Operator and the Monodromy Operator

Once established the well-posedness of (4.2), we can study the qualitative properties of the solutions. Unlike linear autonomous DDEs (2.7), where the state x_t for any $t \geq 0$ may be given by the one-parameter SO-semigroup as seen in Chap. 3, in the linear nonautonomous case we need to deal with two parameters, since also the starting point has to be considered. To this end, we first give a general definition for two-parameter linear and bounded operators $T(t, s) : Y \to Y$ on a general Banach space Y, similarly to what done in Sect. 3.1. Then we go through the case of linear nonautonomous and finally periodic DDEs.

Definition 4.1 (*evolution family*) A family $\{T(t, s)\}_{t \geq s}$ of linear and bounded operators $T(t, s) : Y \to Y$ on a Banach space Y is called an evolution family whenever

- $T(t, t) = I_Y$ for all $t \geq s$;
- $T(t, s) = T(t, v)T(v, s)$ for all $t \geq v \geq s$;
- for any $\varphi \in X$, the function $(t, s) \mapsto T(t, s)\varphi$ is continuous for $t \geq s$.

Definition 4.2 (*evolution operator*) The operator $T(t, s) : X \to X$ associating to the initial function $\varphi \in X$ the state x_t at time $t \geq s$ by (4.2), i.e.,

$$T(t, s)\varphi = x_t(\cdot; s, \varphi), \qquad (4.3)$$

is called the evolution operator.

Proposition 4.1 *The family* $\{T(t, s)\}_{t \geq s}$ *of evolution operators (4.3) defines an evolution family of linear and bounded operators on* X.

Proof It follows the same arguments used in the proof of Proposition 3.1. □

Apart from the need to introduce a two-parameter family, up to this point we have proceeded similarly to the autonomous case. Now we face the main difference: even though the general idea of associating an abstract ODE to an evolution family to follow the evolution of the state in the state space could be appealing, this ODE would result nonautonomous (with a series of difficulties in realizing such association, see [62, Sect. 3.1.1] for a deeper discussion). Hence, a possible corresponding notion of infinitesimal generator would not serve to study the stability (exactly as it happens for finite dimensional linear nonautonomous ODEs like, e.g., (1.3)). On the other hand, by specializing to the periodic case, it is the monodromy operator, together with its spectrum, that becomes central.

Definition 4.3 (*monodromy operator*) Let $T(t, s)$ be the evolution operator (4.3) associated to (4.2) in the periodic case. The operator $U = T(\omega, 0)$ is called the monodromy operator.

Let us note that in the periodic case it also holds

$$T(t + \omega, s) = T(t, s)T(s + \omega, s), \quad t \geq s, \; t, s \in \mathbb{R}. \qquad (4.4)$$

4.1 The Evolution Operator and the Monodromy Operator

Since $\omega > 0$, there exists an integer $q > 0$ such that $q\omega \geq \tau$ and hence $U^q = T(q\omega, 0)$ is compact, by arguments similar to those used in Proposition 3.3. It can also be shown that the spectrum $\sigma(U)$ of U is an at most countable compact set of $\mathbb{C}$ with the only possible accumulation point being zero. Moreover, any element μ in $\sigma(U) \setminus \{0\}$ is an eigenvalue of U which is called a *characteristic* or *Floquet multiplier* of (4.1). Since $\mu \neq 0$, we can express it as $\mu = e^{\omega\lambda}$ and λ is called *characteristic root* or *Floquet exponent*. Being determined modulo $2\pi/\omega$, each μ gives rise to infinitely many λ.

Proposition 4.2 *$\mu \neq 0$ is a characteristic multiplier if and only if there exists $\varphi \neq 0$ such that $T(t + \omega, 0)\varphi = \mu T(t, 0)\varphi$ for all $t \geq 0$.*

Proof It easily follows from (4.4) for $s = 0$. □

The monodromy operator and, therefore, the characteristic multipliers, are defined for $s = 0$. To justify this choice, we prove that they are independent of s.

Proposition 4.3 *If μ is a characteristic multiplier then μ is an eigenvalue of the operator $T(s + \omega, s)$ for all $s \in \mathbb{R}$.*

Proof As for the monodromy operator, for any $s \in \mathbb{R}$ the spectrum of $T(s + \omega, s)$ is an at most countable compact set of $\mathbb{C}$ with zero the only possible accumulation point and any nonzero element of the spectrum is an eigenvalue. We prove that the spectrum of $T(s + \omega, s)$ and $T(t + \omega, t)$ are the same for all $t \neq s$. If $\mu \neq 0$ is an eigenvalue of $T(s + \omega, s)$ then $T(s + \omega, s)\varphi = \mu\varphi$ for some $\varphi \neq 0$. For any $t > s$, we have $T(t + \omega, t)T(t, s)\varphi = T(t + \omega, s)\varphi = T(t, s)T(s + \omega, s)\varphi = \mu T(t, s)\varphi$ and since $T(t, s)\varphi \neq 0$ (otherwise $T(s + \omega, s)^k \varphi$ would be zero for some positive integer k) we have that $\mu \neq 0$ is an eigenvalue of $T(t + \omega, t)$. The thesis follows first by reversing the role of t and s and then by taking $s = 0$. □

The following proposition furnishes a representation of Floquet type of the solutions of (4.2) with initial function in the eigenspace associated to a characteristic multiplier. Moreover, it holds for any finite number of characteristic multipliers.

Proposition 4.4 *Let $\mu = e^{\lambda\omega}$ be a characteristic multiplier with multiplicity $\nu(\mu)$ and let $\Phi = (\varphi_1, \ldots, \varphi_{\nu(\mu)})$ be a basis for the generalized eigenspace $\mathcal{E}(\mu)$. The solutions of (4.2) in the periodic case for $s = 0$ and $\varphi = \Phi b \in \mathcal{E}(\mu)$, with b a $\nu(\mu)$-dimensional vector, are of the form $x_t(\theta; 0, \varphi) = w(t + \theta)e^{C(t+\theta)}b$, $t \geq 0$, $\theta \in [-\tau, 0]$, with C a $\nu(\mu) \times \nu(\mu)$ matrix of single eigenvalue λ and $w(t + \omega) = w(t)$ for all t.*

Proof Since the operator U^q for $q\omega \geq \tau$ is compact, for any μ the generalized eigenspace $\mathcal{E}(\mu)$ is finite dimensional with dimension $\nu(\mu)$. Since $U\mathcal{E}(\mu) \subseteq \mathcal{E}(\mu)$, there exists a matrix B of dimension $\nu(\mu)$ such that $U\Phi = \Phi B$. The only eigenvalue of B is μ. Let $\mu = e^{\lambda\omega}$. We can determine a matrix C with the only eigenvalue λ such that $B = e^{C\omega}$. Let $V(t) = T(t, 0)\Phi e^{-Ct} \in X^{\nu(\mu)}$. Then, for $t \geq 0$, $V(t + \omega) = T(t + \omega, 0)\Phi e^{-C(t+\omega)} = T(t, 0)T(\omega, 0)\Phi e^{-C(t+\omega)} = T(t, 0)\Phi B e^{-C\omega}e^{-Ct} = V(t)$ and $V(t)$ is ω-periodic. Then $T(t, 0)\Phi = V(t)e^{Ct}$. By setting $V(t) = V(t + $

$k\omega$) for $t < 0$ and k such that $t + k\omega > 0$, we can extend V for all $t \in \mathbb{R}$. Every $\varphi \in \mathscr{E}(\mu)$ can be written as $\varphi = \Phi b$, with b a suitable $\nu(\mu)$-dimensional column vector. Then, $x(t + \theta; 0, \varphi) = x_t(\theta; 0, \varphi) = x_{t+\theta}(0; 0, \varphi)$, $\theta \in [-\tau, 0]$, implies $(V(t))(\theta) = (V(t+\theta))(0)e^{C\theta}$. Now $x_t(\theta; 0, \varphi) = w(t+\theta)e^{C(t+\theta)}b$, where $w(t + \theta) = (V(t + \theta))(0)$. □

For linear autonomous DDEs, the stability of the zero solution is determined by the position in $\mathbb{C}$ of the point spectrum of either the solution operator or the associated infinitesimal generator (Theorem 3.4). In the periodic case, this alternative lacks: we can only focus the attention on the spectrum of the monodromy operator. Starting from the Floquet representation of the solution given in Proposition 4.4, a description of the stability of the zero solution follows (see [91, Chap. 8]).

Theorem 4.1 *The zero solution of* (4.1) *is uniformly asymptotically stable if and only if all the characteristic multipliers have modulus* <1 *(or the characteristic exponents have negative real part). It is unstable if there are characteristic multipliers with modulus* >1 *(or characteristic exponents with positive real part).*

Theorem 4.2 *The zero solution of* (4.1) *is uniformly stable if and only if all the characteristic multipliers of* (4.1) *have modulus* ≤ 1 *and if μ is a multiplier with modulus* $=1$ *then all the solutions in the associated eigenspace are bounded.*

In general, the monodromy operator has not an explicit form and the stability analysis needs for suitable approximation techniques to discretize it to compute the characteristic multipliers. In Chap. 6, we introduce a suitable representation of the general evolution operator $T(t, s)$ in (4.3) and we propose a discretization of the latter based on the pseudospectral collocation method. Let us remark again that this approach, which we call the *SO approach* (w.r.t. the *IG approach* proposed in Chap. 5 and relevant to the autonomous case of Chap. 3), is valid for general linear nonautonomous DDEs, not necessarily periodic, and in this context it is presented in Chap. 6. Concerning periodic problems, interested readers can find an overview of alternative methods in [106], see also Sect. 6.4.

Another important observation follows. For linear autonomous DDEs, one can define the characteristic equation (3.13), whose characteristic roots coincide with the eigenvalues of the infinitesimal generator. In general, this cannot be done for linear periodic DDEs but, in some particular cases, a "characteristic equation" whose solutions are the Floquet exponents can be derived. The following example, inspired by [91, Sect. 8.1], represents an instance of periodic DDE in the class (2.8) for which a characteristic equation can be obtained. A similar example, with also distributed delay terms, is treated in Sect. 8.15.

Example 4.1 Consider the linear periodic DDE

$$x'(t) = A(t)x(t) + \sum_{k=1}^{p} B_k(t)x(t - \tau_k), \quad t \geq 0, \qquad (4.5)$$

4.1 The Evolution Operator and the Monodromy Operator

where A and $B_k, k = 1, \ldots, p$, are continuous and ω-periodic $\mathbb{R}^{d \times d}$-valued functions and $\tau_k = n_k \omega$, $n_k \in \mathbb{N}$. By Proposition 4.4 we have that $\mu = e^{\lambda \omega}$ is a characteristic multiplier if and only if there exists a nonzero $\mathbb{R}^d$-valued ω-periodic function w such that $x(t) = w(t)e^{\lambda t}$ satisfies (4.5). Therefore, we get

$$w'(t) = \left(-\lambda I_d + A(t) + \sum_{k=1}^{p} B_k(t) e^{-\lambda \omega n_k} \right) w(t) \qquad (4.6)$$

with $w(t + \omega) = w(t)$. The general solution of (4.6) can be expressed as $w(t) = W(t, 0; \lambda)w(0)$ for $W(t, 0; \lambda)$ the fundamental matrix solution. Note the dependence of W on the characteristic exponent λ. Now, by the periodicity condition $w(\omega) = w(0)$, we can find $w(0) \neq 0$ if and only if λ satisfies the characteristic equation

$$\det(W(\omega, 0; \lambda) - I_d) = 0. \qquad (4.7)$$

For $d = 1$, we have

$$W(t, 0; \lambda) = \exp\left(-\lambda t + \int_0^t A(s)\, rmds + \sum_{k=1}^{p} e^{-\lambda \omega n_k} \int_0^t B_k(s)\, ds \right).$$

Since λ is determined only up to a multiple of $2\pi/\omega$, (4.7) is satisfied if and only if

$$-\lambda + a + \sum_{k=1}^{p} b_k e^{-\lambda \omega n_k} = 0 \qquad (4.8)$$

for $a = \frac{1}{\omega} \int_0^\omega A(s)\, ds$ and $b_k = \frac{1}{\omega} \int_0^\omega B_k(s)\, ds$, $k = 1, \ldots, p$. Observe that (4.8) can be seen as the characteristic equation of $x'(t) = ax(t) + \sum_{k=1}^{p} b_k x(t - \omega n_k)$.

As a concluding remark of this section, let us note that, as already pointed out in Chap. 1, linear periodic ODEs are equivalent to linear autonomous ODEs by means of a periodic nonsingular transformation. This is not true for linear periodic DDEs, as shown in the example proposed in [91, p. 197].

4.2 Linearization and Periodic Solutions

Consider the nonlinear autonomous DDE (2.3) and assume that F satisfies the additional smoothness conditions to ensure that all the forthcoming results are valid.

Definition 4.4 (*periodic solution*) A nonconstant solution $\bar{x}$ of (2.3) on $[-\tau, +\infty)$ is called periodic if there is an $\omega > 0$ such that $\bar{x}(t) = \bar{x}(t + \omega)$ for all $t \geq -\tau$.

Each $\omega > 0$ such that $\bar{x}_\omega = \bar{x}_0$ defines a period of $\bar{x}$. However, there exists a minimal period and all other periods are multiples of the latter.

By following again the underlying idea of the Principle of Linearized Stability, we consider the linearized system at the periodic solution

$$x'(t) = DF(\bar{x}_t)x_t, \quad t \geq 0, \tag{4.9}$$

where $DF(\bar{x}_t)$ is the Fréchet derivative of F at $\bar{x}_t$. The linear operator $DF(\bar{x}_t)$ depends on t and is periodic, i.e., $DF(\bar{x}_{t+\omega}) = DF(\bar{x}_t)$. Hence, the linearized system (4.9) is nonautonomous and periodic. To tell about the solutions closed to the periodic one, we need to state the Principle of Linearized Stability for periodic solutions. It is important to remark that, if $\bar{x}$ is a nonconstant ω-periodic solution of (2.3), the second derivative exists and $\bar{x}''(t) = DF(\bar{x}_t)\bar{x}_t'$. Therefore, $\bar{x}'$ is an ω-periodic solution of (4.9) and then $\bar{x}_0' = U\bar{x}_0'$ for U the associated monodromy operator. Since $\bar{x}' \neq 0$ (for $\bar{x}$ is nonconstant), it follows that $\mu = 1$ is a characteristic multiplier of (4.9). If $\mu = 1$ is simple and it is the only multiplier of modulus equal to one, the periodic solution is called *hyperbolic*. As a consequence, the Principle of Linearized Stability in the periodic case assumes the following form.

Theorem 4.3 *Let $\bar{x}$ be an hyperbolic periodic solution of (2.3). If all the characteristic multipliers of the linearized system (4.9) except $\mu = 1$ are inside the unit circle, then it is exponentially asymptotically stable. If there are characteristic multipliers with modulus larger than one then the periodic solution is unstable.*

It is important to observe that the multiplier $\mu = 1$ is a consequence of the linearization at a periodic solution and it does not affect its stability. This is an essential difference w.r.t. Theorem 3.5 for the solution operator of DDEs linearized at an equilibrium. There we reduce to the linearized equation and to the asymptotic stability of the zero solution by Theorem 3.4. Here this is not possible because the analogous Theorem 4.1 would say that the zero solution of (4.9) is not asymptotically stable.

Let us note that in the literature there are results on existence of periodic solutions for different classes of nonlinear DDEs, for which we have not the analytic expression. In [91, p. 245] "a procedure for determining periodic solutions of some classes of autonomous retarded functional differential equations" is proposed, adding that "the simplest way in which nonconstant periodic solutions of autonomous equations can arise—the so-called Hopf bifurcation—is discussed." In [91, Chap. 11] and [70, Chap. XV] periodic solutions are investigated. In general, such solutions are computed numerically [75–77, 136, 190] and also the study of the linearized equations requires efficient numerical techniques. We close instead this section with an academic example of a nonlinear autonomous DDE with a known periodic solution. This example is resumed in Sect. 8.1.6, showing how the SO approach of Chap. 6 correctly computes the multiplier $\mu = 1$. Let us finally underline that the SO approach works also for linear autonomous DDEs as a particular instance.

4.2 Linearization and Periodic Solutions

Example 4.2 Consider the nonlinear autonomous DDE

$$x'(t) = -\log\left(x\left(t - \frac{\pi}{2}\right)\right)x(t). \tag{4.10}$$

It is easy to verify that $\bar{x}(t) = e^{\sin(t)}$ is a periodic solution of period $\omega = 2\pi$. By following the same procedure of Example 3.1, the linearized equation reads

$$x'(t) = \cos(t)x(t) - e^{\sin(t)+\cos(t)}x\left(t - \frac{\pi}{2}\right), \tag{4.11}$$

which is, indeed, nonautonomous with 2π-periodic coefficients. It is left as an exercise to verify that $\bar{x}'(t)$ solves (4.11) and, therefore, it gives rise to $\mu = 1$ as explained before Theorem 4.3 and as numerically verified in Sect. 8.1.6.

Part II
Numerical Analysis

For readers' convenience, we resume here the basic facts from Part I. By recalling the complexification cited in Remark 3.1, we assume $x(t) \in \mathbb{C}^d$ and the state space $X = C([-\tau, 0], \mathbb{C}^d)$. Then, in Part III, we turn back to $C([-\tau, 0], \mathbb{R}^d)$ for applications.

For $L : X \to \mathbb{C}^d$ a linear and bounded functional, the linear autonomous DDE

$$x'(t) = Lx_t, \ t \in \mathbb{R}, \tag{II.1}$$

can be restated as the abstract linear ODE in the infinite dimensional space X

$$u'(t) = \mathscr{A}u(t), \ t \in \mathbb{R},$$

where $u(t) = x_t$ and $\mathscr{A}$ is the infinitesimal generator of the SO-semigroup $\{T(t)\}_{t \geq 0}$. For $t \geq 0$, the (linear and bounded) solution operator $T(t) : X \to X$ is

$$T(t)\varphi = x_t(\cdot; \varphi), \ \varphi \in X,$$

where $x(\cdot; \varphi)$ is the solution of the Cauchy problem for (II.1) with initial function $x_0 = \varphi$ at time $t = 0$. The (linear and unbounded) infinitesimal generator $\mathscr{A} : \mathscr{D}(\mathscr{A}) \subseteq X \to X$ is

$$\begin{cases} \mathscr{D}(\mathscr{A}) = \{\varphi \in X : \varphi' \in X \text{ and } \varphi'(0) = L\varphi\}, \\ \mathscr{A}\varphi = \varphi'. \end{cases}$$

The eigenvalues of $\mathscr{A}$ are the characteristic roots of (II.1) and we are interested in their numerical computation. We present two approaches: the *IG approach*, based on discretizing the infinitesimal generator, and the *SO approach*, based on discretizing an arbitrary solution operator $T(t)$. In the former, we compute numerical approximations of the eigenvalues of $\mathscr{A}$, giving the characteristic roots directly. In the latter, we compute numerical approximations of the nonzero eigenvalues of $T(t)$: being of type $e^{\lambda t}$ for λ a characteristic root, only $\text{Re}(\lambda)$ can be recovered.

Unlike the IG approach, the SO approach can be used on the more general linear nonautonomous DDEs

$$x'(t) = L(t)x_t, \ t \in \mathbb{I}, \tag{II.2}$$

where, for any $t \in \mathbb{I}$, $L(t) : X \to \mathbb{C}^d$ is a linear and bounded functional. In this more general case, the evolution family $\{T(t, s)\}_{t \geq s}$ replaces the SO-semigroup. For $t \geq s$ in $\mathbb{I}$, the (linear and bounded) evolution operator $T(t, s) : X \to X$ is

$$T(t, s)\varphi = x_t(\cdot; s, \varphi), \ \varphi \in X,$$

where $x(\cdot; s, \varphi)$ is the solution of the Cauchy problem for (II.2) with initial function $x_s = \varphi$ at time $t = s$. The SO approach is used for the numerical computation of the nonzero eigenvalues of an arbitrary evolution operator $T(t, s)$ and, in particular, for the computation of the characteristic multipliers of periodic DDEs, i.e., the eigenvalues of $T(\omega, 0)$ where ω is the (minimal) period of the function $t \mapsto L(t)$. Eventually, we remark that a similar discretization can be used for computing the Lyapunov exponents for the general case (II.2) as done in [49] (see also [32, 83]).

Chapter 5
The Infinitesimal Generator Approach

The IG approach consists in approximating the space X with a finite dimensional linear space X_M, called the *discretization of X of index M*, and the infinitesimal generator $\mathscr{A}$ with a finite dimensional linear operator $\mathscr{A}_M : X_M \to X_M$, called the *discretization of $\mathscr{A}$ of index M*. The index of discretization M is a positive integer such that the larger M, the better the approximations X_M and $\mathscr{A}_M$.

The characteristic roots of (II.1), namely the eigenvalues of $\mathscr{A}$, are then approximated by the eigenvalues of $\mathscr{A}_M$.

In this chapter, we present in Sect. 5.1 a particular method included in this approach, called the *pseudospectral differentiation method of the IG approach*, and in Sect. 5.2 its piecewise version. The convergence analysis for the pseudospectral differentiation method is given in Sect. 5.3 and that for the piecewise version in Sect. 5.4. Other methods based on the IG approach are cited in Sect. 5.5. The implementative aspects of the piecewise version of the pseudospectral differentiation method are considered in Part III.

At first sight, it can seem strange to approximate the infinite spectrum $\sigma(\mathscr{A})$ with the finite spectrum $\sigma(\mathscr{A}_M)$. In the convergence analysis of Sect. 5.3, we explain in which sense the finite spectrum approximates the infinite spectrum. Here, we only anticipate that the larger M, the larger is the number of elements of $\sigma(\mathscr{A})$ that are approximated by elements of $\sigma(\mathscr{A}_M)$ with the closest-to-the origin elements approximated better.

5.1 The Pseudospectral Differentiation Method

The pseudospectral differentiation method of the IG approach consists in particular discretizations of X and $\mathscr{A}$, which are now described. This method first appeared in [38], see also [30].

For any positive integer M, let us introduce the mesh

$$\Omega_M = \{\theta_{M,0}, \theta_{M,1}, \ldots, \theta_{M,M}\} \tag{5.1}$$

on $[-\tau, 0]$, where $0 = \theta_{M,0} > \theta_{M,1} > \cdots > \theta_{M,M} \geq -\tau$. We set $X_M = (\mathbb{C}^d)^{M+1}$ as the discretization of X of index M. An element $\Phi = (\Phi_0, \Phi_1, \ldots, \Phi_M)^T \in X_M$ can be interpreted as a discrete function $\Omega_M \to \mathbb{C}^d$, with $\Phi_m \in \mathbb{C}^d$, $m = 0, 1, \ldots, M$, being the value at $\theta_{M,m}$. Note that the $M+1$ components of Φ are denoted with indices from 0 to M: as we see below, the component of index zero has a special role.

Now, let $P_M : X_M \to X$ be the *discrete Lagrange interpolation operator* associating to any $\Phi \in X_M$ the M-degree $\mathbb{C}^d$-valued polynomial $P_M\Phi \in X$ such that
$$(P_M\Phi)(\theta_{M,m}) = \Phi_m, \quad m = 0, 1, \ldots, M.$$

Here and in the following, for an M-degree $\mathbb{C}^d$-valued polynomial we mean a polynomial $a_M\theta^M + \cdots + a_1\theta + a_0$, $\theta \in [-\tau, 0]$, where $a_M, \ldots, a_1, a_0 \in \mathbb{R}^d$. The discretization $\mathscr{A}_M$ of index M of $\mathscr{A}$ is defined as follows: for any $\Phi \in X_M$, the element $\mathscr{A}_M\Phi \in X_M$ has components

$$\begin{cases} [\mathscr{A}_M\Phi]_0 = LP_M\Phi \\ [\mathscr{A}_M\Phi]_m = (P_M\Phi)'(\theta_{M,m}), \quad m = 1, \ldots, M. \end{cases} \quad (5.2)$$

We can see the approximation $\mathscr{A}_M$ of $\mathscr{A}$ as built in the following way. We start with a discrete function $\Phi \in X_M$, thought as an approximation of a function $\varphi \in \mathscr{D}(\mathscr{A})$ such that
$$\varphi(\theta_{M,m}) = \Phi_m, \quad m = 0, 1, \ldots, M.$$

We want to obtain a discrete function $\mathscr{A}_M\Phi \in X_M$ such that
$$\begin{cases} [\mathscr{A}_M\Phi]_0 \approx \varphi'(\theta_{M,0}) = \varphi'(0) = L\varphi \\ [\mathscr{A}_M\Phi]_m \approx \varphi'(\theta_{M,m}), \quad m = 1, \ldots, M. \end{cases}$$

Thus, we have to approximate the value $L\varphi$, as well as the derivatives of the function φ at the mesh points $\theta_{M,m}$, $m = 1, \ldots, M$, by means of the values Φ_m, $m = 0, 1, \ldots, M$, of φ at the mesh points. An idea is to approximate the function φ by the Lagrange interpolation polynomial $P_M\Phi$, which can be entirely constructed with the values Φ_m, and then set

$$\begin{cases} [\mathscr{A}_M\Phi]_0 = LP_M\Phi \approx L\varphi \\ [\mathscr{A}_M\Phi]_m = (P_M\Phi)'(\theta_{M,m}) \approx \varphi'(\theta_{M,m}), \quad m = 1, \ldots, M. \end{cases}$$

To compute the eigenvalues of $\mathscr{A}_M$, standard software tools require that the linear operator is described as a matrix. Now we find the matrix in $\mathbb{C}^{d(M+1)\times d(M+1)}$ of $\mathscr{A}_M$ relevant to the canonical basis of $\mathbb{C}^{d(M+1)}$. For simplicity of notation, we use always the same symbol for both the operator and the matrix representation.

Given $\Phi \in X_M$, the interpolation polynomial $P_M\Phi$ reads

5.1 The Pseudospectral Differentiation Method

$$(P_M \Phi)(\theta) = \sum_{j=0}^{M} \ell_{M,j}(\theta) \Phi_j, \quad \theta \in [-\tau, 0],$$

where $\ell_{M,0}, \ell_{M,1}, \ldots, \ell_{M,M}$ are the Lagrange coefficients relevant to the nodes of (5.1) given by

$$\ell_{M,j}(\theta) = \prod_{\substack{k=0 \\ k \neq j}}^{M} \frac{\theta - \theta_{M,k}}{\theta_{M,j} - \theta_{M,k}}, \quad \theta \in [-\tau, 0], \ j = 0, 1, \ldots, M.$$

Note that, for $m, j = 0, 1, \ldots, M$,

$$\ell_{M,j}(\theta_{M,m}) = \begin{cases} 1 \text{ if } m = j, \\ 0 \text{ if } m \neq j, \end{cases} \tag{5.3}$$

holds.

By using the previous form of the interpolation polynomial, we can write, for $\Phi \in X_M$,

$$\begin{cases} [\mathscr{A}_M \Phi]_0 = L P_M \Phi = \sum_{j=0}^{M} L\left(\ell_{M,j}(\cdot) I_d\right) \Phi_j \\ [\mathscr{A}_M \Phi]_m = (P_M \Phi)'(\theta_{M,m}) = \sum_{j=0}^{M} \ell'_{M,j}(\theta_{M,m}) \Phi_j, \ m = 1, \ldots, M, \end{cases}$$

where $L\left(\ell_{M,j}(\cdot) I_d\right)$ denotes the $d \times d$ matrix of columns $L\left(\ell_{M,j}(\cdot) e^{(i)}\right)$, $i = 1, \ldots, d$, $e^{(1)}, \ldots, e^{(d)}$ being the canonical vectors of $\mathbb{C}^d$.

We conclude that $\mathscr{A}_M$ is the block matrix given by

$$A_M = \begin{pmatrix} L\left(\ell_{M,0}(\cdot) I_d\right) & L\left(\ell_{M,1}(\cdot) I_d\right) & \cdots & L\left(\ell_{M,M}(\cdot) I_d\right) \\ \ell'_{M,0}(\theta_{M,1}) I_d & \ell'_{M,1}(\theta_{M,1}) I_d & \cdots & \ell'_{M,M}(\theta_{M,1}) I_d \\ \vdots & \vdots & & \vdots \\ \ell'_{M,0}(\theta_{M,M}) I_d & \ell'_{M,1}(\theta_{M,M}) I_d & \cdots & \ell'_{M,M}(\theta_{M,M}) I_d \end{pmatrix}$$

$$= \begin{pmatrix} L\left(\ell_{M,0}(\cdot) I_d\right) & \cdots & L\left(\ell_{M,M}(\cdot) I_d\right) \\ D_M \otimes I_d \end{pmatrix},$$

where $\otimes$ is the Kronecker tensor product [188] (recall that

$$P \otimes Q = \begin{pmatrix} p_{1,1} Q & \cdots & p_{1,v} Q \\ \vdots & \ddots & \vdots \\ p_{u,1} Q & \cdots & p_{u,v} Q \end{pmatrix} \in \mathbb{R}^{ur \times vs}$$

for $P \in \mathbb{R}^{u \times v}$ and $Q \in \mathbb{R}^{r \times s}$). The $M \times (M+1)$ matrix D_M above is

$$D_M = \begin{pmatrix} \ell'_{M,0}(\theta_{M,1}) & \ell'_{M,1}(\theta_{M,1}) & \cdots & \ell'_{M,M}(\theta_{M,1}) \\ \vdots & & & \vdots \\ \ell'_{M,0}(\theta_{M,M}) & \ell'_{M,1}(\theta_{M,M}) & \cdots & \ell'_{M,M}(\theta_{M,M}) \end{pmatrix}.$$

We observe that if we add the first row $\ell'_{M,0}(\theta_{M,0}), \ell'_{M,0}(\theta_{M,1}), \ldots, \ell'_{M,0}(\theta_{M,M})$ to the matrix D_M, we obtain a square matrix that is usually called *differentiation matrix* according to [184]. Therefore, $\mathcal{A}_M$ can be seen as a perturbation of the latter.

Example 5.1 Just for illustration, we observe that for the Hayes Equation (1.4), i.e.,

$$x'(t) = ax(t) + bx(t - \tau),$$

where $a, b \in \mathbb{C}$, for which

$$L\varphi = a\varphi(0) + b\varphi(-\tau), \quad \varphi \in X = C([-\tau, 0], \mathbb{C}),$$

holds, we have

$$A_M = \begin{pmatrix} a\ell_{M,0}(0) + b\ell_{M,0}(-\tau) & \cdots & a\ell_{M,M}(0) + b\ell_{M,M}(-\tau) \\ & D_M & \end{pmatrix}$$
$$= \begin{pmatrix} a + b\ell_{M,0}(-\tau) & b\ell_{M,1}(-\tau) & \cdots & b\ell_{M,M}(-\tau) \\ & D_M & & \end{pmatrix}$$

since $\theta_{M,0} = 0$ and (5.3) holds. Observe that if $\theta_{M,M} = -\tau$, then (5.3) implies

$$A_M = \begin{pmatrix} a & 0 & \cdots & 0 & b \\ & & D_M & & \end{pmatrix}.$$

5.2 The Piecewise Pseudospectral Differentiation Method

In the previous example, we have seen the advantage to take the mesh point $\theta_{M,M}$ coincident with the discrete delay $-\tau$. In applications, when we deal with (2.8), i.e.,

$$x'(t) = Ax(t) + \sum_{k=1}^{p} B_k x(t - \tau_k) + \sum_{k=1}^{p} \int_{-\tau_k}^{-\tau_{k-1}} C_k(\theta) x(t + \theta) \, d\theta, \qquad (5.4)$$

namely the DDE (II.1) with

5.2 The Piecewise Pseudospectral Differentiation Method

$$L\varphi = A\varphi(0) + \sum_{k=1}^{p} B_k \varphi(-\tau_k) + \sum_{k=1}^{p} \int_{-\tau_k}^{-\tau_{k-1}} C_k(\theta)\varphi(\theta)\,d\theta, \quad \varphi \in X,$$

it is quite useful that all the discrete delays $\tau_0, \tau_1, \ldots, \tau_p$ turn out to be mesh points. In order to do this, we introduce the *piecewise version* of the pseudospectral differentiation method. We remark that this version works for the particular DDE (5.4), whereas the non-piecewise version described in the previous section is valid for the general DDE (II.1).

Here, for any positive integer M, we introduce the mesh

$$\Omega_M = \left\{ \theta_{M,0}, \theta_{M,1}^{(1)}, \ldots, \theta_{M,M}^{(1)}, \ldots, \theta_{M,1}^{(p)}, \ldots, \theta_{M,M}^{(p)} \right\}$$

on $[-\tau, 0]$, where $\theta_{M,0} = 0$ and $-\tau_{k-1} > \theta_{M,1}^{(k)} > \cdots > \theta_{M,M}^{(k)} = -\tau_k$, $k = 1, \ldots, p$. Set $X_M = (\mathbb{C}^d)^{pM+1}$ as the discretization of X of index M. An element

$$\Phi = \left(\Phi_0, \Phi_1^{(1)}, \ldots, \Phi_M^{(1)}, \ldots, \Phi_1^{(p)}, \ldots, \Phi_M^{(p)} \right)$$

of X_M is interpreted as a function $\Omega_M \to \mathbb{C}^d$, with Φ_0 being the value at $\theta_{M,0} = 0$ and $\Phi_m^{(k)}, k = 1, \ldots, p, m = 1, \ldots, M$, being the value at $\theta_{M,m}^{(k)}$. We also set

$$\begin{cases} \theta_{M,0}^{(1)} := \theta_{M,0} = 0, \quad \Phi_0^{(1)} := \Phi_0, \\ \theta_{M,0}^{(k)} := \theta_{M,M}^{(k-1)} = -\tau_{k-1}, \quad \Phi_0^{(k)} := \Phi_M^{(k-1)}, \quad k = 2, \ldots, p. \end{cases} \quad (5.5)$$

Now, we introduce the discrete piecewise Lagrange interpolation operator $P_M : X_M \to X$ associating to any $\Phi \in X_M$ the piecewise polynomial $P_M \Phi \in X$ such that, for any $k = 1, \ldots, p$, the restriction $(P_M \Phi)|_{[-\tau_k, -\tau_{k-1}]}$ is the M-degree $\mathbb{C}^d$-valued polynomial such that

$$(P_M \Phi)|_{[-\tau_k, -\tau_{k-1}]}\left(\theta_{M,m}^{(k)}\right) = \Phi_m^{(k)}, \quad m = 0, 1, \ldots, M.$$

Note that, due to the settings (5.5), there are no jumps at the mesh points $-\tau_k$, $k = 1, \ldots, p-1$, and so $P_M \Phi \in X$.

The discretization $\mathscr{A}_M$ of index M of $\mathscr{A}$ is defined as follows: for any $\Phi \in X_M$, the element $\mathscr{A}_M \Phi \in X_M$ has components

$$\begin{cases} [\mathscr{A}_M \Phi]_0 = LP'_M \Phi, \\ [\mathscr{A}_M \Phi]_m^{(k)} = \left((P_M \Phi)|_{[-\tau_k, -\tau_{k-1}]}\right)'\left(\theta_{M,m}^{(k)}\right), \quad k = 1, \ldots, p, \; m = 1, \ldots, M. \end{cases}$$

By introducing, for $k = 1, \ldots, p$, the Lagrange coefficients $\ell_{M,0}^{(k)}, \ell_{M,1}^{(k)}, \ldots, \ell_{M,M}^{(k)}$ relevant to the nodes $\theta_{M,0}^{(k)}, \theta_{M,1}^{(k)}, \ldots, \theta_{M,M}^{(k)}$, we have

$$\begin{cases}
[\mathscr{A}_M \Phi]_0 = L P_M \Phi \\
\quad = A(P_M \Phi)(0) \\
\quad\quad + \sum_{k=1}^{p} B_k (P_M \Phi)(-\tau_k) + \sum_{k=1}^{p} \int_{-\tau_k}^{-\tau_{k-1}} C_k(\theta) (P_M \Phi)(\theta) \, d\theta \\
\quad = A\Phi_0 + \sum_{k=1}^{p} B_k \Phi_M^{(k)} + \sum_{k=1}^{p} \sum_{j=0}^{M} J_j^{(k)} \Phi_j^{(k)} \\
\quad = \left(A + J_0^{(1)}\right) \Phi_0 + \sum_{k=1}^{p-1} \left(\sum_{j=1}^{M-1} J_j^{(k)} \Phi_j^{(k)} + \left(B_k + J_M^{(k)} + J_0^{(k+1)}\right) \Phi_M^{(k)} \right) \\
\quad\quad + \sum_{j=1}^{M-1} J_j^{(p)} \Phi_j^{(p)} + \left(B_p + J_M^{(p)}\right) \Phi_M^{(p)}, \\
[\mathscr{A}_M \Phi]_m^{(k)} = (P_M \Phi)'\left(\theta_{M,m}^{(k)}\right) = \sum_{j=0}^{M} \left(\ell_{M,j}^{(k)}\right)'\left(\theta_{M,m}^{(k)}\right) \Phi_j^{(k)}, \\
k = 1, \ldots, p, \; m = 1, \ldots, M,
\end{cases}$$

where

$$J_j^{(k)} = \int_{-\tau_k}^{-\tau_{k-1}} \ell_{M,j}^{(k)}(\theta) C_k(\theta) \, d\theta, \quad k = 1, \ldots, p, \quad j = 0, 1, \ldots, M.$$

From these equations, one can easily derive the matrix in $\mathbb{C}^{d(pM+1) \times d(pM+1)}$ representing the finite dimensional linear operator $\mathscr{A}_M$ in the canonical basis of $\mathbb{C}^{d(pM+1)}$. The details about the form of this matrix are presented in Chap. 7.

We also observe that, in Part III, we consider a piecewise version of the pseudospectral differentiation method more general than that introduced here. In fact, in Part III, the number of nodes of the mesh Ω_M in the intervals $(-\tau_{k-1}, \tau_k]$, $k = 1, \ldots, p$, is not constant but it depends on k.

5.3 Convergence Analysis

In this section, we are interested in studying how good are the eigenvalues of the discretization $\mathscr{A}_M$, related to the pseudospectral differentiation method, as approximations of the eigenvalues of $\mathscr{A}$. The contents of this section are a detailed reformulation of what has been presented in [38].

The eigenvalues of the operator $\mathscr{A}$ are the complex numbers λ such that

$$\mathscr{A} \varphi = \varphi' = \lambda \varphi \tag{5.6}$$

for some $\varphi \in \mathscr{D}(\mathscr{A}) \setminus \{0\}$.

5.3 Convergence Analysis

Let $\varphi \in X$. Since (5.6) means

$$\varphi(\theta) = \psi(\theta; \lambda, u) := e^{\lambda \theta} u, \quad \theta \in [-\tau, 0], \tag{5.7}$$

for some $u \in \mathbb{C}^d$, we have (5.6) and $\varphi \in \mathscr{D}(\mathscr{A}) \setminus \{0\}$ if and only if

$$\varphi'(0) = \lambda u = L(\psi(\cdot; \lambda, u))$$

for some $u \in \mathbb{C}^d \setminus \{0\}$. Then, we can conclude that λ is an eigenvalue of $\mathscr{A}$ if and only if

$$\det(\lambda I_d - \widehat{L}(\lambda)) = 0, \tag{5.8}$$

where $\widehat{L}(\lambda) : \mathbb{C}^d \to \mathbb{C}^d$ is the linear operator given by

$$\widehat{L}(\lambda) u = L(\psi(\cdot; \lambda, u)), \quad u \in \mathbb{C}^d.$$

Observe that (5.8) is the characteristic Eq. (3.13) of the DDE (II.1) introduced in Proposition 3.4 of Chap. 3.

Now, we show that the eigenvalues of $\mathscr{A}_M$ satisfy a characteristic equation whose form is quite similar to (5.8). Before doing this, we have to study the numerical solution, obtained by a polynomial collocation, of a related problem.

5.3.1 A Related Collocation Problem

For $\lambda \in \mathbb{C}$ and $u \in \mathbb{C}^d$, we denote by $p_M(\cdot; \lambda, u) \in X$ the M-degree collocation polynomial relevant the nodes $\theta_{M,m}$, $m = 0, 1, \ldots, M$, for the problem

$$\begin{cases} \psi'(\theta) = \lambda \psi(\theta), & \theta \in [-\tau, 0], \\ \psi(0) = u, \end{cases} \tag{5.9}$$

whose solution is $\psi(\theta) := \psi(\theta; \lambda, u) = e^{\lambda \theta} u$, $\theta \in [-\tau, 0]$, (recall (5.6) and (5.7)). In other words, $p_M(\cdot; \lambda, u)$ is an M-degree $\mathbb{C}^d$-valued polynomial such that

$$\begin{cases} p'_M(\theta_{M,m}; \lambda, u) = \lambda p_M(\theta_{M,m}; \lambda, u), & m = 1, \ldots, M, \\ p_M(0; \lambda, u) = u. \end{cases} \tag{5.10}$$

Under a condition on the nonzero nodes $\theta_{M,m}$, $m = 1, \ldots, M$, existence and uniqueness of the collocation polynomial, along with an estimate of how well $p_M(\cdot; \lambda, u)$ approximates $\psi(\cdot; \lambda, u)$, is proved in the next proposition. Before presenting it, we introduce the Lagrange interpolation operator $\mathscr{L}_M : X \to X$ associating to any $\varphi \in X$ the $(M-1)$-degree $\mathbb{C}^d$-valued polynomial $\mathscr{L}_M \varphi \in X$ such that

$$(\mathscr{L}_M \varphi)(\theta_{M,m}) = \varphi(\theta_{M,m}), \quad m = 1, \ldots, M.$$

Moreover, let $\ell_{M,1}, \ldots, \ell_{M,M}$ be the Lagrange coefficients relevant to the nodes $\theta_{M,1}, \ldots, \theta_{M,m}$ of (5.1). Finally, we remind that the norm of $\mathscr{L}_M$ is given by

$$\|\mathscr{L}_M\| = \max_{\theta \in [-\tau, 0]} \sum_{j=1}^{M} |\ell_{M,j}(\theta)|$$

and it is known as the *Lebesgue constant* relevant to the nodes $\theta_{M,1}, \ldots \theta_{M,M}$, see, e.g., [186] and the references therein or [184].

Proposition 5.1 *Let B be an open bounded subset of $\mathbb{C}$. If the sequence of meshes $\{\Omega_M\}$ is such that*

$$\lim_{M \to \infty} \frac{\|\mathscr{L}_M\|}{M} = 0, \tag{5.11}$$

then there exists a positive integer $M_0(B)$ such that, for any index $M \geq M_0(B)$, $\lambda \in B$ and $u \in \mathbb{C}^d$, there exists a unique collocation polynomial $p_M(\cdot; \lambda, u)$ for (5.9) and

$$\|p_M(\cdot; \lambda, u) - \psi(\cdot; \lambda, u)\|_X \leq C_2(B) \cdot \frac{C_1(B)^M}{M!} \cdot \|u\|_\infty, \tag{5.12}$$

where

$$C_1(B) = \max_{z \in \overline{B}} |z| \, \tau$$

and

$$C_2(B) = 2 \cdot \max_{z \in \overline{B}} \left(1 + |z| \frac{e^{\operatorname{Re}(z)\tau} - 1}{\operatorname{Re}(z)}\right) \cdot \max\left\{e^{\operatorname{Re}(z)\tau}, 1\right\},$$

$\overline{B}$ being the closure of B. Moreover, the linear operator $\mathbb{C}^d \to X$ given by

$$u \mapsto p_M(\cdot; \lambda, u), \quad u \in \mathbb{C}^d, \tag{5.13}$$

is an holomorphic function of $\lambda \in B$.

Proof Let $\lambda \in \mathbb{C}$ and let $u \in \mathbb{C}^d$. By introducing the linear Volterra integral operator $\mathscr{K} : X \to X$ given by

$$(\mathscr{K}\varphi)(\theta) = \int_0^\theta \varphi(\theta) \, d\theta, \quad \varphi \in X, \ \theta \in [-\tau, 0],$$

we can rewrite (5.9) and (5.10) as equations in the space X: (5.9) becomes

5.3 Convergence Analysis

$$\psi = u + \lambda \mathcal{K} \psi$$

and (5.10) becomes

$$p_M = u + \lambda \mathcal{K} \mathcal{L}_M p_M, \tag{5.14}$$

where we set $\psi = \psi(\cdot; \lambda, u)$ and $p_M = p_M(\cdot; \lambda, u)$ and u denotes the constant function in X of value $u \in \mathbb{C}^d$. In fact, we can write

$$\psi(\theta) = u + \int_0^\theta \psi'(\theta) \, d\theta = u + \int_0^\theta \lambda \psi(\theta) \, d\theta, \quad \theta \in [-\tau, 0],$$

instead of (5.9) and

$$p_M(\theta) = u + \int_0^\theta p'_M(\theta) \, d\theta = u + \int_0^\theta (\mathcal{L}_M \lambda p_M)(\theta) \, d\theta, \quad \theta \in [-\tau, 0],$$

instead of (5.10), since the first row of (5.10) says that $p'_M = \mathcal{L}_M \lambda p_M$.

By introducing the error $e_M := p_M - \psi$, we obtain the following equation for e_M:

$$e_M = \lambda \mathcal{K} \mathcal{L}_M e_M + \lambda \mathcal{K} r_M, \tag{5.15}$$

where $r_M := \mathcal{L}_M \psi - \psi$. We have that $e_M \in X$ is a solution of (5.15) if and only if $e_M = \lambda \mathcal{K} \widehat{e}_M$, where $\widehat{e}_M \in X$ satisfies the equation

$$\widehat{e}_M = \lambda \mathcal{L}_M \mathcal{K} \widehat{e}_M + r_M. \tag{5.16}$$

The reason for introducing the Eq. (5.16) is that the operators $\mathcal{K}$ and $\mathcal{L}_M$ appear here in the reverse order with respect to (5.15): this fact permits to prove existence and uniqueness of a solution for (5.16) in an easy way. Here is the proof of the equivalence of (5.15) and (5.16). If e_M is a solution of (5.15), then, for $\widehat{e}_M = \mathcal{L}_M e_M + r_M$, we have $e_M = \lambda \mathcal{K} (\mathcal{L}_M e_M + r_M) = \lambda \mathcal{K} \widehat{e}_M$ and $\widehat{e}_M = \mathcal{L}_M e_M + r_M = \lambda \mathcal{L}_M \mathcal{K} \widehat{e}_M + r_M$. Vice-versa, if $\widehat{e}_M \in X$ is a solution of (5.16), then, by setting $e_M = \lambda \mathcal{K} \widehat{e}_M$, we have $\widehat{e}_M = \lambda \mathcal{L}_M \mathcal{K} \widehat{e}_M + r_M = \mathcal{L}_M e_M + r_M$ and so $e_M = \lambda \mathcal{K} (\mathcal{L}_M e_M + r_M) = \lambda \mathcal{K} \mathcal{L}_M e_M + \lambda \mathcal{K} r_M$.

Now, we prove that if (5.11) holds, then there exists a positive integer $M_0(B)$ such that (5.16), rewritten as

$$(I_X - \lambda \mathcal{L}_M \mathcal{K}) \widehat{e}_M = r_M, \tag{5.17}$$

has a unique solution, for $M \geq M_0(B)$ and $\lambda \in B$.

We begin by observing that the operator $I_X - \lambda \mathcal{K}$ is invertible with inverse

$$\left[(I_X - \lambda \mathscr{K})^{-1} \varphi\right](\theta) = \varphi(\theta) + \lambda \int_0^\theta e^{\lambda(\theta-s)} \varphi(s)\, ds, \ \varphi \in X, \ \theta \in [-\tau, 0], \tag{5.18}$$

and so

$$\left\|(I_X - \lambda \mathscr{K})^{-1}\right\| \leq 1 + |\lambda| \frac{e^{\operatorname{Re}(\lambda)\tau} - 1}{\operatorname{Re}(\lambda)}.$$

Note that (5.18) follows by rewriting $(I_X - \lambda \mathscr{K})\psi = \varphi$, where $\varphi, \psi \in X$, as $\psi - \varphi = \lambda \mathscr{K}(\psi - \varphi) + \lambda \mathscr{K}\varphi$, i.e., $(\psi - \varphi)' = \lambda(\psi - \varphi) + \lambda \varphi$ and $(\psi - \varphi)(0) = 0$.

Now, we consider the operator $I_X - \lambda \mathscr{L}_M \mathscr{K}$ in the left-hand side of (5.17) as a perturbation of the operator $I_X - \lambda \mathscr{K}$: we have

$$I_X - \lambda \mathscr{L}_M \mathscr{K} = I_X - \lambda \mathscr{K} - \lambda (\mathscr{L}_M - I_X) \mathscr{K}. \tag{5.19}$$

The Banach Perturbation Lemma (see, e.g., [124, Theorem 10.1]) says that if Y is a Banach space and $A, E : Y \to Y$ are linear bounded operators such that A is invertible and

$$\|E\| < \frac{1}{\|A^{-1}\|},$$

then $A + E$ is invertible and

$$\left\|(A + E)^{-1}\right\| \leq \frac{\|A^{-1}\|}{1 - \|E\|\|A^{-1}\|}.$$

We want to apply it with $A = I_X - \lambda \mathscr{K}$ and $E = -\lambda(\mathscr{L}_M - I_X)\mathscr{K}$.

A well-known result from interpolation theory (a Jackson's theorem, see [163, Corollary 1.4.1]) says that, for a function $\varphi \in X$, the following bound

$$\|\mathscr{L}_M \varphi - \varphi\|_X \leq (1 + \|\mathscr{L}_M\|) \cdot 6\omega \left(\varphi; \frac{\tau}{2(M-1)}\right) \tag{5.20}$$

holds, where

$$\omega(\varphi; \delta) = \sup_{\substack{\theta_1, \theta_2 \in [-\tau, 0] \\ |\theta_1 - \theta_2| \leq \delta}} \|\varphi(\theta_1) - \varphi(\theta_2)\|_\infty, \ \delta > 0,$$

is the modulus of continuity of φ. As a consequence, for the norm of the perturbation $\lambda(\mathscr{L}_M - I_X)\mathscr{K}$ in (5.19), we have

$$\|\lambda(\mathscr{L}_M - I_X)\mathscr{K}\| \leq |\lambda|(1 + \|\mathscr{L}_M\|)\frac{3\tau}{M-1}.$$

5.3 Convergence Analysis

This is the point that explains the importance to have $\mathscr{L}_M \mathscr{K}$ instead of $\mathscr{K} \mathscr{L}_M$ in the equation: in this manner, we have the interpolation error operator $\mathscr{L}_M - I_X$ as applied to functions smoothed by the integral operator $\mathscr{K}$.

Under the hypothesis (5.11), there exists a positive integer $M_0(B)$ such that, for $M \geq M_0(B)$ and $\lambda \in B$, we have

$$\|\lambda (\mathscr{L}_M - I_X) \mathscr{K}\| \cdot \left\|(I_X - \lambda \mathscr{K})^{-1}\right\|$$
$$\leq |\lambda|(1 + \|\mathscr{L}_M\|) \frac{3\tau}{M-1} \cdot \left(1 + |\lambda| \frac{e^{\operatorname{Re}(\lambda)\tau} - 1}{\operatorname{Re}(\lambda)}\right)$$
$$\leq (1 + \|\mathscr{L}_M\|) \frac{3\tau}{M-1} \cdot \max_{z \in B} |z| \left(1 + |z| \frac{e^{\operatorname{Re}(z)\tau} - 1}{\operatorname{Re}(z)}\right) \leq \frac{1}{2}$$

and then, by the Banach Perturbation Lemma, we can conclude that $I_X - \lambda \mathscr{L}_M \mathscr{K}$ is invertible with

$$\left\|(I_X - \lambda \mathscr{L}_M \mathscr{K})^{-1}\right\| \leq 2 \left\|(I_X - \lambda \mathscr{K})^{-1}\right\| \leq 2 \left(1 + |\lambda| \frac{e^{\operatorname{Re}(\lambda)\tau} - 1}{\operatorname{Re}(\lambda)}\right).$$

Therefore, for $M \geq M_0(B)$, the Eq. (5.16) has a unique solution $\widehat{e}_M$ with

$$\|\widehat{e}_M\|_X \leq \left\|(I_X - \lambda \mathscr{L}_M \mathscr{K})^{-1}\right\| \|r_M\|_X \leq 2 \left(1 + |\lambda| \frac{e^{\operatorname{Re}(\lambda)\tau} - 1}{\operatorname{Re}(\lambda)}\right) \|r_M\|_X.$$

Thus, (5.15) has a unique solution e_M and, then, there exists a unique collocation polynomial p_M and

$$\|e_M\|_X \leq \|\lambda \mathscr{K}\| \|\widehat{e}_M\|_X \leq 2 |\lambda| \tau \left(1 + |\lambda| \frac{e^{\operatorname{Re}(\lambda)\tau} - 1}{\operatorname{Re}(\lambda)}\right) \|r_M\|_X. \quad (5.21)$$

Another well-known result from interpolation theory (the Cauchy Interpolation Remainder [67, Theorem 3.1.1]) says that

$$\|r_M\|_X = \|\mathscr{L}_M \psi - \psi\|_X \leq \frac{\tau^M \|\psi^{(M)}\|_X}{M!} \leq \frac{\tau^M |\lambda|^M \max\{e^{\operatorname{Re}(\lambda)\tau}, 1\} \|u\|_\infty}{M!}. \quad (5.22)$$

By (5.21) and (5.22), we obtain the bound (5.12).

As for the holomorphicity of (5.13) on B, it is sufficient to observe that we can express the collocation polynomial $p_M(\cdot; \lambda, u)$ as

$$p_M(\cdot; \lambda, u) = \psi(\cdot; \lambda, u) + e_M = \psi(\cdot; \lambda, u) + \lambda \mathscr{K} \widehat{e}_M$$
$$= \psi(\cdot; \lambda, u) + \lambda \mathscr{K} (I_X - \lambda \mathscr{L}_M \mathscr{K})^{-1} (\mathscr{L}_M - I_X) \psi(\cdot; \lambda, u).$$

The holomorphicity of the linear operator (5.13) follows by that of the linear operators $u \mapsto \psi(\cdot; \lambda, u)$ and $u \mapsto \lambda \mathscr{K} (I_X - \lambda \mathscr{L}_M \mathscr{K})^{-1} (\mathscr{L}_M - I_X) \psi(\cdot; \lambda, u)$, $u \in \mathbb{C}^d$. □

We observe that (5.11) is fulfilled by *Chebyshev zeros* (or type-I nodes, [184])

$$\theta_{M,m} = \frac{\tau}{2}\left(\cos\left(\frac{2m-1}{2M}\right)\pi - 1\right), \quad m = 1, \ldots, M, \qquad (5.23)$$

for which $\|\mathscr{L}_M\| = O(\log(M))$, $M \to \infty$, holds.

Moreover, we remark that (5.22) is valid for any choice of $\theta_{M,1}, \ldots, \theta_{M,M}$ and it can be improved for particular choices of such nodes. The best is obtained in the case of (5.23). In this case, we have, instead of (5.22),

$$\|r_M\|_X \leq \frac{\tau^M |\lambda|^M \max\left\{e^{\mathrm{Re}(\lambda)\tau}, 1\right\} \|u\|_\infty}{2^{M-1} M!}.$$

As a consequence we can improve the estimate (5.12) to

$$\|p_M(\cdot; \lambda, u) - \psi(\cdot; \lambda, u)\|_X \leq 2C_2(B) \cdot \frac{\left(\frac{C_1(B)}{2}\right)^M}{M!} \cdot \|u\|_\infty, \qquad (5.24)$$

which is only a little bit better than (5.12).

Observe that, for $M \geq M_0(B)$ and $\lambda \in B$, we have $p_M(\cdot; \lambda, 0) = 0$. This follows by observing that the zero polynomial satisfies (5.10) for $u = 0$.

Now, we are in position to introduce the characteristic equation for the discretization $\mathscr{A}_M$, whose form is quite similar to (5.8).

Proposition 5.2 *Let B be an open bounded subset of $\mathbb{C}$ and let $M \geq M_0(B)$, where $M_0(B)$ is given in Proposition 5.1. The eigenvalues of $\mathscr{A}_M$ in B are the complex numbers $\lambda \in B$ such that*

$$\det\left(\lambda I_d - \widehat{L}_M(\lambda)\right) = 0, \qquad (5.25)$$

where $\widehat{L}_M(\lambda) : \mathbb{C}^d \to \mathbb{C}^d$ is the linear operator given by

$$\widehat{L}_M(\lambda)u = L\left(p_M(\cdot; \lambda, u)\right), \quad u \in \mathbb{C}^d.$$

Proof The eigenvalues of $\mathscr{A}_M$ in B are the complex numbers $\lambda \in B$ such that

$$\mathscr{A}_M \Phi = \lambda \Phi \qquad (5.26)$$

for some $\Phi \in X_M \setminus \{0\}$. Let $\Phi \in X_M$. Note that (5.26) reads

5.3 Convergence Analysis

$$\begin{cases} [\mathscr{A}_M \Phi]_0 = L P_M \Phi = \lambda \Phi_0, \\ [\mathscr{A}_M \Phi]_m = (P_M \Phi)' (\theta_{M,m}) = \lambda \Phi_m = \lambda (P_M \Phi) (\theta_{M,m}), \quad m = 1, \ldots, M. \end{cases} \quad (5.27)$$

The second equations in (5.27) say that, for some $u \in \mathbb{C}^d$,

$$P_M \Phi = p_M (\cdot; \lambda, u). \tag{5.28}$$

Assume that (5.26) holds, i.e., (5.28) and the first of (5.27) hold, for some $\Phi \in X_M \setminus \{0\}$. Note that $u \neq 0$. In fact, if $u = 0$, then $P_M \Phi = p_M (\cdot; \lambda, 0) = 0$ and so

$$\Phi = \left((P_M \Phi) (\theta_{M,0}), (P_M \Phi) (\theta_{M,1}), \ldots, (P_M \Phi) (\theta_{M,M}) \right) = 0.$$

Since $L p_M(\cdot; \lambda, u) = L P_M \Phi = \lambda \Phi_0 = \lambda u$ holds and $u \neq 0$, we obtain (5.25).

Vice-versa, if (5.25) holds, i.e., $L p_M(\cdot; \lambda, u) = \lambda u$ holds for some $u \in \mathbb{C}^d \setminus \{0\}$, then we have (5.28) and the first of (5.27) with

$$\Phi = \left(p_M (\theta_{M,0}; \lambda, u), p_M (\theta_{M,1}; \lambda, u), \ldots, p_M (\theta_{M,M}; \lambda, u) \right).$$

Note that $\Phi_0 = p_M (\theta_{M,0}; \lambda, u) = u \neq 0$ and then $\Phi \neq 0$. We conclude that (5.26) holds for some $\Phi \in X_M \setminus \{0\}$. □

5.3.2 Convergence of the Eigenvalues

In order to estimate how good are the eigenvalues of $\mathscr{A}_M$ as approximations of the characteristic roots of (II.1), we have to compare the continuous characteristic Eq. (5.8)

$$\det \left(\lambda I_d - \widehat{L}(\lambda) \right) = 0$$

with the discrete characteristic Eq. (5.25)

$$\det \left(\lambda I_d - \widehat{L}_M(\lambda) \right) = 0.$$

A comparison between these two equations can be done by means of the Rouché's Theorem from Complex Analysis (see, e.g., [161, Sect. 7.7]). This theorem states that if complex-valued functions f and g are holomorphic inside and on some closed simple contour K and $|g(z) - f(z)| < |f(z)|$, $z \in K$, holds, then f and g have the same number of zeros inside K, where each zero is counted as many times as its multiplicity.

Let λ^* be a characteristic root of (II.1) with multiplicity ν^*. Note that the multiplicity ν^* of the root λ^* as a solution of the characteristic equation is equal to the algebraic multiplicity $\nu(\lambda^*)$ of λ^* as an eigenvalue of $\mathscr{A}$. Let B be an open ball in $\mathbb{C}$ of center λ^* and radius r_0^* such that λ^* is the unique characteristic root of (II.1) in B. The existence of such a ball is guaranteed from the fact that the characteristic roots of (II.1) are isolated (see Sect. 3.2). We also assume that (5.11) holds.

Consider the complex-valued functions

$$f(\lambda) := \det\left(\lambda I_d - \widehat{L}(\lambda)\right), \quad \lambda \in B,$$

and

$$g_M(\lambda) := \det\left(\lambda I_d - \widehat{L}_M(\lambda)\right), \quad \lambda \in B.$$

Clearly, f is an holomorphic function on B. The holomorphicity of g_M on B follows by the holomorphicity of the linear operator (5.13). The Rouché's Theorem says that if

$$|g_M(\lambda) - f(\lambda)| < |f(\lambda)|, \quad \left|\lambda - \lambda^*\right| = r^*,$$

where $0 < r^* < r_0^*$, then the discrete characteristic equation

$$g_M(\lambda) = \det\left(\lambda I_d - \widehat{L}_M(\lambda)\right) = 0$$

has ν^* roots in $B(\lambda^*, r^*)$ by counting multiplicities. Here $B(\lambda^*, r)$ is the open ball of center λ^* and radius r^*. Of course, the error of these discrete characteristic roots as approximations of λ^* is not larger in modulus than r^*. The following proposition gives an estimate of $|g_M(\lambda) - f(\lambda)|$ for $\lambda \in B$.

Proposition 5.3 *There exists a positive integer $M_1(B)$ with $M_1(B) \geq M_0(B)$, $M_0(B)$ given in Proposition 5.1, such that, for any index $M \geq M_1(B)$, we have*

$$|g_M(\lambda) - f(\lambda)| \leq C_3(B) \cdot C_2(B) \cdot \frac{C_1(B)^M}{M!}, \quad \lambda \in B,$$

where $C_1(B)$ and $C_2(B)$ are defined in Proposition 5.1 and

$$C_3(B) = \max_{\substack{z \in \overline{B} \\ \Delta \in \mathbb{C}^{d \times d} \\ \|\Delta\| \leq 1}} \left\|(\det)'\left(z I_d - \widehat{L}(z) + \Delta\right)\right\| \cdot \|L\|,$$

with $(\det)' : \mathbb{C}^{d \times d} \to \left(\mathbb{C}^{d \times d} \to \mathbb{R}\right)$ the derivative of $\det : \mathbb{C}^{d \times d} \to \mathbb{R}$.

Proof By (5.12), we obtain, for $\lambda \in B$,

$$\left\|\left(\widehat{L}_M(\lambda) - \widehat{L}(\lambda)\right) u\right\|_\infty = \|L\left(p_M(\cdot; \lambda, u) - \psi(\cdot; \lambda, u)\right)\|$$
$$\leq \|L\| \cdot \|p_M(\cdot; \lambda, u) - \psi(\cdot; \lambda, u)\|_X$$
$$\leq \|L\| \cdot C_2(B) \cdot \frac{C_1(B)^M}{M!} \cdot \|u\|_\infty, \quad u \in \mathbb{C}^d,$$

and then

$$\left\|\widehat{L}_M(\lambda) - \widehat{L}(\lambda)\right\| \leq \|L\| \cdot C_2(B) \cdot \frac{C_1(B)^M}{M!}.$$

5.3 Convergence Analysis

Now, let $M_1(B)$ be a positive integer such that $M_1(B) \geq M_0(B)$ and

$$\|L\| \cdot C_2(B) \cdot \frac{C_1(B)^M}{M!} \leq 1, \quad M \geq M_1(B).$$

For $M \geq M_1(B)$, we have

$$\begin{aligned}
|g_M(\lambda) - f(\lambda)| &= \left|\det\left(\lambda I_d - \widehat{L}_M(\lambda)\right) - \det\left(\lambda I_d - \widehat{L}(\lambda)\right)\right| \\
&= \left|\int_0^1 (\det')(\lambda I_d - \widehat{L}(\lambda) + s(\widehat{L}(\lambda) - \widehat{L}_M(\lambda))(\widehat{L}(\lambda) - \widehat{L}_M(\lambda))\, ds\right| \\
&\leq \max_{s \in [0,1]} \left\|(\det')(\lambda I_d - \widehat{L}(\lambda) + s(\widehat{L}(\lambda) - \widehat{L}_M(\lambda)))\right\| \cdot \left\|\widehat{L}(\lambda) - \widehat{L}_M(\lambda)\right\| \\
&\leq \max_{\substack{z \in \overline{B} \\ \Delta \in \mathbb{C}^{d \times d} \\ \|\Delta\| \leq 1}} \left\|(\det)'\left(zI_d - \widehat{L}(z) + \Delta\right)\right\| \cdot \left\|\widehat{L}_M(\lambda) - \widehat{L}(\lambda)\right\|.
\end{aligned}$$

The thesis follows. $\square$

Assume $M \geq M_1(B)$. For $\lambda \in B$, a Taylor expansion gives

$$\left|f(\lambda) - \frac{f^{(\nu^*)}(\lambda^*)}{\nu^*!}(\lambda - \lambda^*)^{\nu^*}\right| \leq \frac{1}{(\nu^* + 1)!} \max_{z \in \overline{B}}\left|f^{(\nu^* + 1)}(z)\right| |\lambda - \lambda^*|^{\nu^* + 1}.$$

By choosing $r \in (0, r_0^*)$ such that

$$\frac{1}{(\nu^* + 1)!} \max_{z \in \overline{B}}\left|f^{(\nu^* + 1)}(z)\right| |\lambda - \lambda^*|^{\nu^* + 1} < \frac{1}{2} \cdot \frac{\left|f^{(\nu^*)}(\lambda^*)\right|}{\nu^*!} |\lambda - \lambda^*|^{\nu^*}, \quad |\lambda - \lambda^*| = r,$$

and this is obtained for

$$r < r_1^* := \frac{1}{2} \cdot \frac{(\nu^* + 1)\left|f^{(\nu^*)}(\lambda^*)\right|}{\max_{z \in \overline{B}}\left|f^{(\nu^* + 1)}(z)\right|},$$

we have

$$|f(\lambda)| > \frac{1}{2} \cdot \frac{\left|f^{(\nu^*)}(\lambda^*)\right|}{\nu^*!} |\lambda - \lambda^*|^{\nu}, \quad |\lambda - \lambda^*| = r.$$

By Proposition 5.3, we conclude that, for $0 < r < \min\{r_0^*, r_1^*\}$, we have

$$|g_M(\lambda) - f(\lambda)| < |f(\lambda)|, \quad |\lambda - \lambda^*| = r,$$

if
$$C_3(B) \cdot C_2(B) \cdot \frac{C_1(B)^M}{M!} \leq \frac{1}{2} \cdot \frac{\left|f^{(\nu)}(\lambda^*)\right|}{\nu^*!} r^{\nu^*}. \tag{5.29}$$

Now, by
$$\lim_{M \to \infty} \frac{C_1(B)^M}{M!} = 0,$$

we have that there exists a positive integer $M_2(B)$ with $M_2(B) \geq M_1(B)$ such that, for any index $M \geq M_2(B)$,

$$r^* := \left(\frac{C_3(B) \cdot C_2(B) \cdot \frac{C_1(B)^M}{M!}}{\frac{1}{2} \cdot \frac{\left|f^{(\nu^*)}(\lambda^*)\right|}{\nu^*!}} \right)^{\frac{1}{\nu^*}} < \min\{r_0^*, r_1^*\}.$$

Thus, for $M \geq M_2(B)$, since (5.29) is true with equality for $r = r^*$, we obtain
$$|g_M(\lambda) - f(\lambda)| < |f(\lambda)|, \quad |\lambda - \lambda^*| = r^*,$$

and then we can conclude that there exist ν^* discrete characteristic roots in $B(\lambda^*, r^*)$. Therefore, we have proved the following convergence theorem.

Theorem 5.1 *Let λ^* be a characteristic root of (II.1) with multiplicity ν^* and let B be an open ball of center λ^* such that λ^* is the unique characteristic root of (II.1) in B. Assume that the nodes $\theta_{M,m}$, $m = 1, \ldots, M$, satisfy (5.11). Then, there exists a positive integer $M_2(B)$ such that, for any index $M \geq M_2(B)$, there exists ν^* discrete characteristic roots (i.e., eigenvalues of $\mathscr{A}_M$) $\lambda_{M,1}^*, \ldots, \lambda_{M,\nu^*}^*$, each counted with its multiplicity, such that*

$$\max_{i=1,\ldots,\nu^*} \left|\lambda_{M,i}^* - \lambda^*\right| \leq \left(\frac{C_3(B) \cdot C_2(B) \cdot \frac{C_1(B)^M}{M!}}{\frac{1}{2} \cdot \frac{\left|f^{(\nu^*)}(\lambda^*)\right|}{\nu^*!}} \right)^{\frac{1}{\nu^*}}, \tag{5.30}$$

where $C_1(B)$ and $C_2(B)$ are defined in Proposition 5.1 and $C_3(B)$ is defined in Proposition 5.3.

As an immediate corollary, we obtain what follows.

Theorem 5.2 *Assume that the open ball $B(0, R)$ of center 0 and radius R in $\mathbb{C}$ contains the characteristic roots $\lambda_1^*, \ldots, \lambda_K^*$ of (II.1) with multiplicities $\nu_1^*, \ldots, \nu_K^*$, respectively. Assume that the nodes $\theta_{M,m}$, $m = 1, \ldots, M$, satisfy (5.11). Then, there exists a positive integer $M_3(R)$ such that, for any index $M \geq M_3(R)$ and for each $k = 1, \ldots, K$, there exists ν_k^* discrete characteristic roots $\lambda_{M,k,1}^*, \ldots, \lambda_{M,k,\nu_k^*}^*$, each counted with its multiplicity, such that*

5.3 Convergence Analysis

$$\max_{\substack{k=1,\ldots,K \\ i=1,\ldots,v_k}} \left|\lambda^*_{M,k,i} - \lambda^*\right| = O\left(\left(\frac{(\tau R)^M}{M!}\right)^{\frac{1}{v^*}}\right), \quad M \to \infty. \quad (5.31)$$

where $v^* = \max\{v_1^*, \ldots, v_K^*\}$.

Proof For any $k = 1, \ldots, K$, we consider an open ball B_k of center λ_k^* contained in $B(0, R)$ and such that λ_k^* is the unique characteristic root of (II.1) in B_k. Moreover, for any $k = 1, \ldots, K$, let M_k be a positive integer such that, for $M \geq M_k$, there exists v_k^* discrete characteristic roots $\lambda^*_{M,k,1}, \ldots, \lambda^*_{M,k,v_k^*}$ counted with multiplicities such that

$$\max_{i=1,\ldots,v_k^*} \left|\lambda^*_{M,k,i} - \lambda_k^*\right| \leq \left(\frac{C_3(B_k) \cdot C_2(B_k) \cdot \frac{C_1(B_k)^M}{M!}}{\frac{1}{2} \cdot \frac{\left|f^{(v_k^*)}(\lambda_k^*)\right|}{v_k^*!}}\right)^{\frac{1}{v_k^*}}$$

$$\leq \left(\frac{C_3(B_k) \cdot C_2(B_k) \cdot \frac{(\tau R)^M}{M!}}{\frac{1}{2} \cdot \frac{\left|f^{(v_k^*)}(\lambda_k^*)\right|}{v_k^*!}}\right)^{\frac{1}{v_k^*}}.$$

Now, let $M_3 = M_3(B)$ be a positive integer such that $M_3 \geq \max_{k=1,\ldots,K} M_k$ and

$$\frac{(\tau R)^M}{M!} \leq 1, \quad M \geq M_3.$$

Then, for $M \geq M_3$, we have

$$\max_{\substack{k=1,\ldots,K \\ i=1,\ldots,v_k^*}} \left|\lambda^*_{M,k,i} - \lambda_k^*\right| \leq \max_{k=1,\ldots,K} \left(\frac{C_3(B_k) \cdot C_2(B_k)}{\frac{1}{2} \cdot \frac{\left|f^{(v_k^*)}(\lambda_k^*)\right|}{v_k^*!}}\right)^{\frac{1}{v_k^*}} \cdot \max_{k=1,\ldots,K} \left(\frac{(\tau R)^M}{M!}\right)^{\frac{1}{v_k^*}}$$

$$\leq \max_{k=1,\ldots,K} \left(\frac{C_3(B_k) \cdot C_2(B_k)}{\frac{1}{2} \cdot \frac{\left|f^{(v_k^*)}(\lambda_k^*)\right|}{v_k^*!}}\right)^{\frac{1}{v_k^*}} \cdot \left(\frac{(\tau R)^M}{M!}\right)^{\frac{1}{v^*}}.$$

□

The previous theorem explains in which sense the finite spectrum $\sigma(\mathscr{A}_M)$ approximates the infinite spectrum $\sigma(\mathscr{A})$. As it has been anticipated at the beginning of this chapter, we see by (5.31) that the closest-to-the origin characteristic roots are approximated better.

The convergence

$$O\left(\left(\frac{(\tau R)^M}{M!}\right)^{\frac{1}{\nu^*}}\right), \quad M \to \infty, \tag{5.32}$$

of the discrete characteristic roots to the continuous characteristic roots in the open ball of center 0 and radius R is of *infinite order*, i.e., it is faster of any convergence of type $O\left(M^{-p}\right)$, $p > 0$. The fact that the discrete characteristic roots have the convergence (5.32) is often referred to in the literature by saying that they have *spectral accuracy* (see, e.g., [184]).

As already observed at the end of the Sect. 5.3.1, the assumption (5.11) on the nodes $\theta_{M,m}$, $m = 1, \ldots, M$, is satisfied by (5.23). By using (5.24) instead of (5.12), we see that, for such nodes, the convergence order is slightly better than (5.32):

$$O\left(\left(\frac{\left(\frac{\tau R}{2}\right)^M}{M!}\right)^{\frac{1}{\nu^*}}\right), \quad M \to \infty.$$

5.3.3 Quadrature for Distributed Delays

In our previous convergence analysis, we have assumed that the values of the functional L can be exactly computed. However, in case of a distributed delay, the integral involved has to be approximated by a quadrature rule in general. Thus, a more refined convergence analysis should consider the situation where, for any index of discretization M, the linear functional $L : X \to \mathbb{C}^d$ is approximated by a linear functional $L_M : X \to \mathbb{C}^d$, whose values can be exactly computed. If, indeed, the values of L can be exactly computed, as in the case of discrete delays only, we have $L_M = L$ for any index M. When the approximations L_M are considered, we have to replace L with L_M in the definition (5.2) of the discretization $\mathscr{A}_M$. Then, Proposition 5.2 is still valid with

$$\widehat{L}_M(\lambda)u = L_M\left(p_M\left(\cdot; \lambda, u\right)\right), \quad u \in \mathbb{C}^d. \tag{5.33}$$

Given an open bounded subset B of $\mathbb{C}$, under the assumptions that there exists $D > 0$ such that $\|L_M\| \leq D$ for any index M and that

$$E_M(B) := \max_{\substack{z \in \overline{B} \\ u \in \mathbb{C}^d \\ \|u\|_\infty = 1}} \|L_M\left(\psi\left(\cdot; z, u\right)\right) - L\left(\psi\left(\cdot; z, u\right)\right)\|_\infty \to 0, \quad M \to \infty,$$

with $\psi\left(\cdot; z, u\right)$ defined in (5.7), the following proposition replaces Proposition 5.3.

5.3 Convergence Analysis

Proposition 5.4 *There exists a positive integer $M_1(B)$ with $M_1(B) \geq M_0(B)$, $M_0(B)$ given in Proposition 5.1, such that, for any index $M \geq M_1(B)$, we have*

$$|g_M(\lambda) - f(\lambda)| \leq C_3(B) \cdot C_2(B) \cdot \frac{C_1(B)^M}{M!} + C_4(B) \cdot E_M(B), \quad \lambda \in B,$$

where $C_1(B)$ and $C_2(B)$ are defined in Proposition 5.1 and

$$C_3(B) = C_4(B) \cdot D, \tag{5.34}$$

$$C_4(B) = \max_{\substack{z \in B \\ \Delta \in \mathbb{C}^{d \times d} \\ \|\Delta\| \leq 1}} \left\| (\det)' \left(z I_d - \widehat{L}(z) + \Delta \right) \right\|$$

with $\det'$ the derivative of $\det$.

As a consequence, Theorem 5.1 has, instead of (5.30), the estimate

$$\max_{i=1,\ldots,v^*} \left| \lambda_{M,i}^* - \lambda^* \right| \leq \left(\frac{C_3(B) \cdot C_2(B) \cdot \frac{C_1(B)^M}{M!} + C_4(B) \cdot E_M(B)}{\frac{1}{2} \cdot \frac{\left| f^{(v^*)}(\lambda^*) \right|}{v^*!}} \right)^{\frac{1}{v^*}}$$

(where $C_3(B)$ is now given in (5.34)) and Theorem 5.2 has the estimate

$$\max_{\substack{k=1,\ldots,K \\ i=1,\ldots,v_k}} \left| \lambda_{M,k,i}^* - \lambda^* \right| = O\left(\left(\frac{(\tau R)^M}{M!} \right)^{\frac{1}{v^*}} \right) + O\left((E_M(B(0,R)))^{\frac{1}{v^*}} \right), \quad M \to \infty,$$

instead of (5.31). Note that, in case of a distributed delay

$$L\varphi = \int_{-\tau}^{0} w(\theta) \varphi(\theta) \, d\theta, \quad \varphi \in X,$$

$E_M(B)$ is the maximum error when the quadrature rule is applied to the integrals

$$L\psi(\cdot; z, u) = \int_{-\tau}^{0} w(\theta) e^{\lambda \theta} u \, d\theta, \quad z \in \overline{B}, \quad u \in \mathbb{C}^d, \quad \|u\|_{\infty} = 1.$$

5.4 Convergence of the Piecewise Method

The convergence analysis for the piecewise pseudospectral differentiation method follows the same line of the non-piecewise case. Now, instead of the polynomial $p_M(\cdot;\lambda,u)$ of Sect. 5.3.1, we consider the piecewise polynomial $p_M(\cdot;\lambda,u)$ such that, for any $k=1,\ldots,p$, the restriction $p_M(\cdot;\lambda,u)|_{[-\tau_k,-\tau_{k-1}]}$ is an M-degree $\mathbb{C}^d$-valued polynomial and such that

$$\begin{cases} (p_M|_{[-\tau_k,-\tau_{k-1}]})'\left(\theta_{M,m}^{(k)};\lambda,u\right) = \lambda p_M\left(\theta_{M,m}^{(k)};\lambda,u\right), & k=1,\ldots,p, \quad m=1,\ldots,M, \\ p_M(0;\lambda,u) = u. \end{cases}$$

Now, in (5.14), $\mathscr{L}_M$ is the piecewise Lagrange interpolation operator $\mathscr{L}_M : X \to X$ associating to any $\varphi \in X$ the piecewise polynomial $\mathscr{L}_M\varphi \in X$ such that, for $k=1,\ldots,p$, $(\mathscr{L}_M\varphi)|_{[-\tau_k,-\tau_{k-1}]}$ is the $(M-1)$-degree $\mathbb{C}^d$-valued polynomial such that

$$(\mathscr{L}_M\varphi)|_{[-\tau_k,-\tau_{k-1}]}\left(\theta_{M,m}^{(k)}\right) = \varphi\left(\theta_{M,m}^{(k)}\right), \quad k=1,\ldots,p, \quad m=1,\ldots,M.$$

All proceed as in the non-piecewise case and all the estimates presented for the non-piecewise case are still valid: in Theorem 5.2, τ now is tha maximum difference $\tau_k - \tau_{k-1}$ for $k=1,\ldots,p$. Of course, in the condition (5.11) we have

$$\|\mathscr{L}_M\| = \max_{k=1,\ldots,p}\left(\max_{\theta\in[-\tau,0]}\sum_{j=1}^M \left|\ell_{M,j}^{(k)}(\theta)\right|\right),$$

where $\ell_{M,j}^{(k)},\ldots,\ell_{M,j}^{(k)}$ are the Lagrange coefficients relevant to the nodes $\theta_{M,1}^{(k)},\ldots,\theta_{M,m}^{(k)}$.

In the implementation of the piecewise pseudospectral differentiation method of Part III, we use the *extremal Chebyshev nodes* (or type-II nodes)

$$\theta_{M,m}^{(k)} = \frac{\tau_k - \tau_{k-1}}{2}\cos\left(\frac{m\pi}{M}\right) - \frac{\tau_k + \tau_{k-1}}{2}, \quad m=0,1,\ldots,M,$$

on each interval $[\tau_{k-1},\tau_k]$, $k=1,\ldots,p$. Such nodes satisfy the condition (5.11).

5.5 Other Methods

In the previous sections, we have presented the pseudospectral differentiation method of the IG approach along with its piecewise version. These methods correspond to discretize the space X and the operator $\mathscr{A}$ in a particular manner.

5.5 Other Methods

Of course, there are other ways to do this. Since $\mathscr{A}$ is a differentiation operator (recall that $\mathscr{A}\varphi = \varphi'$ for $\varphi \in \mathscr{D}(\mathscr{A})$), any manner to discretize the derivative can provide, in principle, a method of the IG approach for computing the characteristic roots of (II.1).

For example, given a function $y : [t_0, t_0 + h] \to \mathbb{C}^d$ and nodes

$$t_i = t_0 + c_i h, \quad i = 1, \ldots, s,$$

where $c_1 \ldots, c_s \in [0, 1]$ are distinct, one can approximate the derivatives $y'(t_i), i = 1, \ldots, s$, in terms of the values $y(t_i), i = 0, 1, \ldots, s$, by means of a Runge-Kutta (RK) method (see, e.g., [55, 90]). In fact, given a RK method with abscissae c_i, $i = 1, \ldots, s$, and coefficients $a_{ij}, i, j = 1, \ldots, s$, by replacing in the RK equations

$$Y_i = y(t_0) + h \sum_{j=1}^{s} a_{ij} K_j, \quad i = 1, \ldots, s,$$

Y_i with $y(t_i)$ and by solving for the unknows $K_j, j = 1, \ldots, s$, we obtain approximations of the derivatives $y'(t_i), i = 1, \ldots, s$. By using this idea, one can construct a discretization of $\mathscr{A}$ based on RK methods. This discretization has been presented in [30, 37].

It is also possible (see [28] or, again, [30]) to construct a discretization of $\mathscr{A}$ based on linear multistep methods [55, 90], instead of RK methods.

However, both the methods cited above provide only a finite order of convergence of the discrete characteristic roots to the continuous characteristic roots: we obtain, for a continuous characteristic root with multiplicity v^*, a convergence

$$O\left(h^{\frac{p}{v^*}}\right), \quad h \to 0,$$

where h is a submultiple of τ and p is the order of convergnece of the underlying method. For this reason, the pseudospectral differentiation method, which has spectral accuracy, should be preferred.

Chapter 6
The Solution Operator Approach

In the SO approach, the eigenvalues of an evolution operator

$$T := T(s+h, s) : X \to X$$

for the DDE (II.2) as defined in (4.1), where $s \in \mathbb{R}$ and $h > 0$, are approximated by the eigenvalues of a finite dimensional approximation of T. In order to obtain such an approximation, it is convenient to express T in a suitable form.

Besides the space $X = C([-\tau, 0], \mathbb{C}^d)$, we introduce the space $X^+ := C([0, h], \mathbb{C}^d)$ equipped with the maximum norm $\|z\|_{X^+} = \max_{t \in [0,h]} \|z(t)\|_\infty$, $z \in X^+$, the space $X^\pm := C([-\tau, h], \mathbb{C}^d)$ (which does not need a norm) and the map $V : X \times X^+ \to X^\pm$ given by

$$(V(\varphi, z))(\theta) = \begin{cases} \varphi(0) + \int_0^\theta z(t)\, dt & \text{if } \theta \in [0, h] \\ \varphi(\theta) & \text{if } \theta \in [-\tau, 0] \end{cases} \quad (6.1)$$

for $(\varphi, z) \in X \times X^+$ and $\theta \in [-r, h]$. Observe that $V(\varphi, z)$ prolongs φ from $[-\tau, 0]$ to $[-\tau, h]$ by using the solution of the differential equation

$$\begin{cases} v'(t) = z(t), & t \in [0, h], \\ v(0) = \varphi(0). \end{cases}$$

Finally, we introduce the linear operator $\mathscr{F}_s : X^\pm \to X^+$ defined by

$$(\mathscr{F}_s v)(t) = L(s+t)v_t, \quad v \in X^\pm, \quad t \in [0, h]. \quad (6.2)$$

By using the map V, we can express the evolution operator T in the form

$$T\varphi = V(\varphi, z^*)_h, \quad \varphi \in X, \quad (6.3)$$

where $z^* \in X^+$ is the unique solution of the fixed point equation

$$z^* = \mathscr{F}_s V(\varphi, z^*) \tag{6.4}$$

in the space X^+. In fact, z^* is the shift $t \mapsto x'(s+t)$, $t \in [0, h]$, of the derivative x', where x is the solution of (4.2).

The SO approach consists in approximating the spaces X and X^+ with finite dimensional linear spaces X_M and X_N^+, called the *discretizations of X of index M* and *of X^+ of index N*, and the evolution operator T with a finite dimensional linear operator $T_{M,N} : X_M \to X_M$, called the *discretization of T of indices M and N*. The eigenvalues of T are then approximated by the eigenvalues of $T_{M,N}$.

In this chapter, we present in Sect. 6.1 a particular method included in this approach, called the *pseudospectral collocation method of the SO approach*. In Sect. 6.2 we study the collocation equation that implicitely defines such a method. The convergence analysis is accomplished in Sect. 6.3. Other methods based on the SO approach are cited in Sect. 5.5. The implementative aspects of the SO approach with the pseudospectral collocation method are presented in Part III.

6.1 The Pseudospectral Collocation Method

The pseudospectral collocation method of the SO approach consists in particular discretizations of X, X^+ and T, which are now described. We introduce the discretizations of the spaces X and X^+ along with restriction and prolongation linear operators, which are then used in the construction of the discretization of T. This method first appeared in [44] and the contents of the following sections are a detailed reformulation of the presentation in that paper.

6.1.1 Discretization of X

We treat separately the two cases $h \geq \tau$ and $h < \tau$ (recall that τ is the maximum delay of the equation).

6.1.1.1 The Case $h \geq \tau$

For a given positive integer M, we introduce the mesh in $[-\tau, 0]$

$$\Omega_M := \{\theta_{M,0}, \theta_{M,1}, \ldots, \theta_{M,M}\},$$

6.1 The Pseudospectral Collocation Method

where $0 \geq \theta_{M,0} > \theta_{M,1} > \cdots > \theta_{M,M} \geq -\tau$ and set $X_M := \left(\mathbb{C}^d\right)^{M+1}$ as the discretization of the space X. This is very similar to the discretization of X introduced for the pseudospectral differentiation method of the IG approach as in Chap. 5, but here $\theta_{M,0} = 0$ is no longer required. An element $\Phi \in X_M$ is written as $\Phi = (\Phi_0, \Phi_1, \ldots, \Phi_M)^T$, where $\Phi_m \in \mathbb{C}^d$, $m = 0, 1, \ldots, M$.

The *restriction operator* $R_M : X \to X_M$ is given by

$$R_M \varphi = (\varphi(\theta_{M,0}), \varphi(\theta_{M,1}), \ldots, \varphi(\theta_{M,M}))^T, \quad \varphi \in X.$$

In other words, the restriction R_M makes discrete a continuous function $\varphi \in X$ by considering its values at the nodes $\theta_{M,0}, \theta_{M,1}, \ldots, \theta_{M,M}$.

The *prolongation operator* $P_M : X_M \to X$ is the discrete Lagrange interpolation operator:

$$(P_M \Phi)(\theta) = \sum_{m=0}^{M} \ell_{M,m}(\theta) \Phi_m, \quad \Phi \in X_M, \ \theta \in [-\tau, 0],$$

where $\ell_{M,0}, \ell_{M,1}, \ldots, \ell_{M,m}$, are the Lagrange coefficients relevant to the nodes of Ω_M. In other words, the prolongation P_M makes continuous a discrete function $\Phi \in X_M$ by interpolating the values $\Phi_0, \Phi_1, \ldots, \Phi_M$ at the nodes $\theta_{M,0}, \theta_{M,1}, \ldots, \theta_{M,M}$ with an M-degree $\mathbb{C}^d$-valued polynomial.

We observe that

$$R_M P_M = I_{X_M} \tag{6.5}$$

and

$$P_M R_M = \mathscr{L}_M, \tag{6.6}$$

where $\mathscr{L}_M : X \to X$ is the Lagrange interpolation operator that associates to a function $\varphi \in X$ the M-degree $\mathbb{C}^d$-valued polynomial $\mathscr{L}_M \varphi$ such that $(\mathscr{L}_M \varphi)(\theta_{M,m}) = \varphi(\theta_{M,m})$, $m = 0, 1, \ldots, M$.

6.1.1.2 The Case $h < \tau$

We adopt the same type of discretization as in the previous case $h \geq \tau$, but now it is made in a piecewise manner on the successive intervals $[-h, 0], [-2h, -h], \ldots$

Let Q be the minimum positive integer q such that $qh \geq \tau$. Note that $Q > 1$. We set $\theta^{(q)} = -qh$, $q = 0, \ldots, Q-1$, and $\theta^{(Q)} = -\tau$. For a given positive integer M, we consider the mesh in $[-\tau, 0]$

$$\Omega_M := \bigcup_{q=1}^{Q} \left\{\theta_{M,0}^{(q)}, \ldots, \theta_{M,M}^{(q)}\right\},$$

where

$$\begin{cases} 0 = \theta^{(0)} \geq \theta^{(1)}_{M,0} > \cdots > \theta^{(1)}_{M,M} = \theta^{(1)}, \\ \theta^{(q-1)} = \theta^{(q)}_{M,0} > \cdots > \theta^{(q)}_{M,M} = \theta^{(q)}, \quad q = 2, \ldots, Q-1, \\ \theta^{(Q-1)} = \theta^{(Q)}_{M,0} > \cdots > \theta^{(Q)}_{M,M} \geq \theta^{(Q)} = -\tau, \end{cases}$$

and set $X_M := (\mathbb{C}^d)^{QM+1}$ as the discretization of X. This discretization is similar to the discretization introduced for the piecewise pseudospectral differentiation method in Sect. 5.2, but here it is related to h and not to the discrete delays of the equation.

An element $\Phi \in X_M$ is written as $\Phi = \left(\Phi_0^{(1)}, \ldots, \Phi_{M-1}^{(1)}, \ldots, \Phi_0^{(Q)}, \ldots, \Phi_{M-1}^{(Q)}, \Phi_M^{(Q)} \right)^T$, where $\Phi_m^{(q)} \in \mathbb{C}^d$, $q = 1, \ldots, Q$ and $m = 0, 1, \ldots, M-1$, and $\Phi_M^{(Q)} \in \mathbb{C}^d$. We also set $\Phi_M^{(q)} := \Phi_0^{(q+1)}$, $q = 1, \ldots, Q-1$.

The restriction operator $R_M : X \to X_M$ is given by $R_M \varphi = \Phi$, $\varphi \in X$, where $\Phi_m^{(q)} = \varphi(\theta_{M,m}^{(q)})$, $q = 1, \ldots, Q$, $m = 0, 1, \ldots, M$.

The prolongation operator $P_M : X_M \to X$ is the discrete piecewise Lagrange interpolation operator

$$(P_M \Phi)(\theta) = \sum_{m=0}^{M} \ell_{M,m}^{(q)}(\theta) \Phi_m^{(q)}, \quad \Phi \in X_M, \ \theta \in [\theta^{(q)}, \theta^{(q-1)}], \ q = 1, \ldots, Q,$$

where, for $q = 1, \ldots, Q$, $\ell_{M,0}^{(q)}, \ell_{M,1}^{(q)}, \ldots, \ell_{M,M}^{(q)}$ are the Lagrange coefficients relevant to the nodes $\theta_{M,0}^{(q)}, \theta_{M,1}^{(q)}, \ldots, \theta_{M,M}^{(q)}$.

The relations (6.5) and (6.6) are still valid, but $\mathscr{L}_M$ is now the piecewise Lagrange interpolation operator that associates to each function $\varphi \in X$ the piecewise polynomial $\mathscr{L}_M \varphi$ such that, for any $q = 1, \ldots, Q$, the restriction $(\mathscr{L}_M \varphi)|_{[\theta^{(q)}, \theta^{(q-1)}]}$ is the M-degree $\mathbb{C}^d$-valued polynomial with values $\varphi\left(\theta_{M,m}^{(q)}\right)$ at the nodes $\theta_{M,m}^{(q)}$, $m = 0, 1, \ldots, M$.

6.1.2 Discretization of X^+

The space X^+ is discretized in the same way as the space X in the case $h \geq \tau$. For a given positive integer N, let

$$\Omega_N^+ := \{t_{N,1}, \ldots, t_{N,N}\}$$

be a mesh in $[0, h]$, where $0 \leq t_{N,1} < \cdots < t_{N,N} \leq h$ and set $X_N^+ := (\mathbb{C}^d)^N$ as the discretization of X^+. An element $Z \in X_N^+$ is written as $Z = (Z_1, \ldots, Z_N)$, where $Z_n \in \mathbb{C}^d$, $n = 1, \ldots, N$. Observe that, unlike the indices for the nodes in Ω_M and

6.1 The Pseudospectral Collocation Method

the elements in X_M, which start from 0, the indices for the nodes in Ω_N^+ and the elements in X_N^+ start from 1.

The restriction operator $R_N^+ : X^+ \to X_N^+$ and the prolongation operator $P_N^+ : X_N^+ \to X^+$ are given by

$$R_N^+ z = (z(t_{N,1}), \ldots, z(t_{N,N}))^T, \quad z \in X^+,$$

and

$$(P_N^+ Z)(t) = \sum_{n=1}^{N} \ell_{N,n}^+(t) Z_n, \quad Z \in X_N^+, \ t \in [0, h],$$

where $\ell_{N,1}^+, \ldots, \ell_{N,n}^+$ are the Lagrange coefficients relevant to the nodes $t_{N,1}, \ldots, t_{N,N}$. We have

$$R_N^+ P_N^+ = I_{X_N^+} \tag{6.7}$$

and

$$P_N^+ R_N^+ = \mathscr{L}_N^+, \tag{6.8}$$

where $\mathscr{L}_N^+ : X^+ \to X^+$ is the Lagrange interpolation operator that associates to each function $z \in X^+$ the $(N-1)$-degree $\mathbb{C}^d$-valued polynomial $\mathscr{L}_N^+ z$ such that $(\mathscr{L}_N^+ z)(t_{N,n}) = z(t_{N,n})$, $n = 1, \ldots, N$.

6.1.3 Discretization of T

For given positive integers M and N, we consider as the discretization of indices M and N of the evolution operator T the finite dimensional operator $T_{M,N} : X_M \to X_M$ given by

$$T_{M,N}\Phi = R_M V(P_M \Phi, P_N^+ Z_N^*)_h, \quad \Phi \in X_M, \tag{6.9}$$

where $Z_N^* \in X_N^+$ is a solution of the fixed point equation

$$Z^* = R_N^+ \mathscr{F}_s V(P_M \Phi, P_N^+ Z^*) \tag{6.10}$$

in the space X_N^+.

Observe that (6.9) and (6.10) are the discrete counterparts of (6.3) and (6.4), respectively. In particular, we have that the functions $\varphi \in X$ and $z^* \in X^+$ in (6.3) and (6.4) are replaced in (6.9) and (6.10) with the interpolating polynomials $P_M \Phi$ and

$P_N^+ Z^*$ and the Eq. (6.4) is discretized by a collocation at the nodes $t_{N,1}, \ldots, t_{N,N}$. In fact, it can be particularized as $Z_n^* = \mathscr{F}_s V(P_M \phi, P_N^+ Z^*)(t_{N,n})$, $n = 1, \ldots, N$.

The matrix of $T_{M,N}$ relevant to the canonical basis of X_M is recovered in Part III. In the next section, we study the collocation Eq. (6.10). Indeed, we consider a more general equation, where $P_M \Phi$ is replaced with a generic function $\phi \in X$.

6.2 The Collocation Equation

Consider the collocation equation

$$Z^* = R_N^+ \mathscr{F}_s V(\phi, P_N^+ Z^*), \tag{6.11}$$

where $\phi \in X$. In this section, we show that (6.11) has a unique solution Z_N^* and compare Z_N^* to the unique solution z^* of

$$z^* = \mathscr{F}_s V(\phi, z^*). \tag{6.12}$$

We begin by observing that we can decompose the map V introduced in (6.1) as

$$V(\varphi, z) = V_1 \varphi + V_2 z, \quad (\varphi, z) \in X \times X^+, \tag{6.13}$$

where $V_1 : X \to X^\pm$ and $V_2 : X \to X^\pm$ are linear operators given by

$$V_1 \varphi = V(\varphi, 0), \quad \varphi \in X, \tag{6.14}$$

and

$$V_2 z = V(0, z), \quad z \in X^+. \tag{6.15}$$

By using the decomposition (6.13), we can rewrite (6.11) as

$$(I_{X_N^+} - R_N^+ \mathscr{F}_s V_2 P_N^+) Z^* = R_N^+ \mathscr{F}_s V_1 \phi. \tag{6.16}$$

As a first step, we establish the following lemma, which reduces the invertibility of the operator

$$I_{X_N^+} - R_N^+ \mathscr{F}_s V_2 P_N^+ : X_N^+ \to X_N^+ \tag{6.17}$$

in (6.16) to the invertibility of the operator

$$I_{X^+} - \mathscr{L}_N^+ \mathscr{F}_s V_2 : X^+ \to X^+. \tag{6.18}$$

6.2 The Collocation Equation

Proposition 6.1 *If the operator* (6.18) *is invertible, then the operator* (6.17) *is invertible. Moreover, given* $Z \in X_N^+$, *the unique solution* w_N^* *of the equation*

$$\left(I_{X^+} - \mathcal{L}_N^+ \mathcal{F}_s V_2\right) w^* = P_N^+ Z \tag{6.19}$$

in X^+ *and the unique solution* Z_N^* *of the equation*

$$\left(I_{X_N^+} - R_N^+ \mathcal{F}_s V_2 P_N^+\right) Z^* = Z, \tag{6.20}$$

in X_N^+ *are related by* $Z_N^* = R_N^+ w_N^*$ *and* $w_N^* = P_N^+ Z_N^*$.

Proof Assume that the operator (6.18) is invertible. For the unique solution w_N^* of (6.19) we have, by (6.8), $w_N^* = P_N^+ (R_N^+ \mathcal{F}_s V_2 w_N^* + Z)$ and so, by (6.7), $R_N^+ w_N^* = R_N^+ \mathcal{F}_s V_2 w_N^* + Z$ and $R_N^+ w_N^* = R_N^+ \mathcal{F}_s V_2 w_N^* + Z = R_N^+ \mathcal{F}_s V_2 P_N^+ R_N^+ w_N^* + Z$, i.e., $R_N^+ w_N^*$ is a solution of (6.20).

Vice-versa, if Z^* is a solution of (6.20), then, by (6.8), $P_N^+ Z^* = \mathcal{L}_N^+ \mathcal{F}_s V_2 P_N^+ Z^* + P_N^+ Z$, i.e., $P_N^+ Z^*$ is a solution of (6.19). Therefore, if Z_1^* and Z_2^* are two solutions of (6.20), then $P_N^+ Z_1^* = w_N^* = P_N^+ Z_2^*$, where w_N^* is the unique solution of (6.19). Then, by (6.7), $Z_1^* = R_N^+ P_N^+ Z_1^* = R_N^+ P_N^+ Z_2^* = Z_2^*$. We conclude that $Z_N^* := R_N^+ w_N^*$ is the unique solution of (6.20). This proves the invertibility of the operator (6.17). Moreover, we have $w_N^* = P_N^+ Z_N^*$, since $P_N^+ Z_N^*$ is a solution of (6.19). □

As a consequence of the previous proposition, we have that, under the assumption of invertibility for the operator (6.18), the Eq. (6.11) has a unique solution Z_N^* and $Z_N^* = R_N^+ w_N^*$ and $w_N^* = P_N^+ Z_N^*$ hold, where w_N^* is the unique solution of

$$w^* = \mathcal{L}_N^+ \mathcal{F}_s V(\phi, w^*). \tag{6.21}$$

In fact, the Eq. (6.21) can be rewritten as $(I_{X_N} - \mathcal{L}_N^+ \mathcal{F}_s V_2)w^* = \mathcal{L}_N^+ \mathcal{F}_s V_1 \phi = P_N^+ Z$, where $Z = R_N^+ \mathcal{F}_s V_1 \phi$ and (6.11) can be rewritten as (6.16), namely $(I_{X_N^+} - R_N^+ \mathcal{F}_s V_2 P_N^+) Z^* = R_N^+ \mathcal{F}_s V_1 \phi = Z$.

In the next Theorem 6.1, we show that the operator (6.18) is invertible under suitable assumptions and give also a bound for

$$\left\| w_N^* - z^* \right\|_{X^+} = \left\| P_N^+ Z_N^* - z^* \right\|_{X^+},$$

where z^* is the unique solution of (6.12). Before presenting such theorem, we introduce the subspace X_{Lip}^+ of X^+ of the Lipschitz continuous functions of X^+, which is necessary for convergence (see Assumption (C1)). Such subspace X_{Lip}^+ is equipped with the norm

$$\|z\|_{X_{\text{Lip}}^+} = \text{Lip}(z) + \|z\|_{X^+}, \quad z \in X_{\text{Lip}}^+,$$

where $\text{Lip}(z)$ denotes the Lipschitz constant of z.

Theorem 6.1 *If*

(C1) *the operator* $\mathscr{F}_s V_2 : X^+ \to X^+$ *has range contained in* X^+_{Lip} *and* $\mathscr{F}_s V_2 : X^+ \to X^+_{\text{Lip}}$ *is bounded;*

(C2) *the sequence of meshes* $\{\Omega^+_N\}$ *is such that*

$$\lim_{N\to\infty} \frac{\|\mathscr{L}^+_N\|}{N} = 0,$$

where

$$\|\mathscr{L}^+_N\| = \max_{t\in[0,h]} \sum_{n=1}^{N} |\ell_{N,n}(t)|$$

is the Lebesgue constant relevant to the nodes $t_{N,1},\ldots,t_{N,N}$; *then, there exists a positive integer* N_0 *such that, for any index* $N \geq N_0$, *the operator* $I_{X^+} - \mathscr{L}^+_N \mathscr{F}_s V_2$ *is invertible with*

$$\left\|(I_{X^+} - \mathscr{L}^+_N \mathscr{F}_s V_2)^{-1}\right\| \leq 2 \left\|(I_{X^+} - \mathscr{F}_s V_2)^{-1}\right\|. \tag{6.22}$$

In addition, (6.21) *has a unique solution* w^*_N *and*

$$\|w^*_N - z^*\|^+_{X^+} \leq 2 \left\|(I_{X^+} - \mathscr{F}_s V_2)^{-1}\right\| \|\mathscr{L}^+_N z^* - z^*\|_{X^+} \tag{6.23}$$

holds, where z^* *is the solution of* (6.12).

Proof We begin by observing that the operator $I_{X^+} - \mathscr{F}_s V_2$ is invertible. This follows by noticing that the equation $(I_{X^+} - \mathscr{F}_s V_2)z = g$, where $g \in X^+$, has a unique solution $z \in X^+$ if and only if

$$\begin{cases} x'(t) = L(s+t)x_t + g(t), & t \in [0,h], \\ x_s = 0 \end{cases}$$

has a unique solution. In fact, we can rewrite the latter as $x' = \mathscr{F}_s V_2 x' + g$.

The operator $I_{X^+} - \mathscr{L}^+_N \mathscr{F}_s V_2$ can be considered as a perturbation of the bounded and invertible operator $I_{X^+} - \mathscr{F}_s V_2$. In fact: $I_{X^+} - \mathscr{L}^+_N \mathscr{F}_s V_2 = I_{X^+} - \mathscr{F}_s V_2 - (\mathscr{L}^+_N - I_{X^+})\mathscr{F}_s V_2$.

Now, we use the interpolation error bound (5.20) of Chap. 5, but adapted to the interval $[0,h]$. For a function $z \in X^+$, we have

$$\|\mathscr{L}^+_N z - z\|_{X^+} \leq (1 + \|\mathscr{L}^+_N\|) 6\omega\left(z; \frac{h}{2(N-1)}\right), \tag{6.24}$$

6.2 The Collocation Equation

where

$$\omega(z; \delta) = \sup_{\substack{t_1, t_2 \in [0, h] \\ |t_1 - t_2| \leq h}} \|z(t_1) - z)(t_2)\|_\infty, \quad \delta > 0,$$

is the modulus of continuity of z. Since (C2) holds, (6.24) yields

$$\left\|(\mathscr{L}_N^+ - I_{X^+})|_{X_{\text{Lip}}^+}\right\|_{X^+ \leftarrow X_{\text{Lip}}^+} \to 0, \quad N \to \infty,$$

and then, since (C1) holds, we have

$$\left\|(\mathscr{L}_N^+ - I_{X^+})\mathscr{F}_s V_2\right\| \leq \left\|(\mathscr{L}_N^+ - I_{X^+})|_{X_{\text{Lip}}^+}\right\|_{X^+ \leftarrow X_{\text{Lip}}^+} \|\mathscr{F}_s V_2\|_{X_{\text{Lip}}^+ \leftarrow X^+} \to 0, \quad N \to \infty.$$

Now, let N_0 be a positive integer such that

$$\left\|(\mathscr{L}_N^+ - I_{X^+})\mathscr{F}_s V_2\right\| \cdot \left\|(I_{X^+} - \mathscr{F}_s V_2)^{-1}\right\| \leq \frac{1}{2}, \quad N \geq N_0.$$

By the Banach Perturbation Lemma ([124, Theorem 10.1], as used in Chap. 5), as applied with $A = I_{X^+} - \mathscr{F}_s V_2$ and $E = -(\mathscr{L}_N^+ - I_{X^+})$, we obtain, for any index $N \geq N_0$, the invertibility of $I_{X^+} - \mathscr{L}_N^+ \mathscr{F}_s V_2$ and the bound (6.22).

It remains to prove the bound (6.23). By setting $w_N^* = z^* + e_N^*$ in (6.21), we have the equation $(I_{X^+} - \mathscr{L}_N^+ \mathscr{F}_s V_2) e_N^* = \mathscr{L}_N^+ z^* - z^*$ for the error e_N^*. It is obtained by using the property $V(\varphi, z_1 + z_2) = V(\varphi, z_1) + V_2 z_2$, $\varphi \in X$, $z_1, z_2 \in X^+$, of the map V and the fact that z^* solves (6.12). The bound (6.23) now follows from the bound (6.22). □

We have seen in Chap. 2 that the linear nonautonomous functional $L(t) : X \to \mathbb{C}^d$ can be expressed as

$$L(t)\varphi = \int_{-\tau}^{0} d_\theta[\eta(t, \theta)]\varphi(\theta), \quad \varphi \in X,$$

where $\eta(t, \cdot) \in NBV\left([-\tau, 0], \mathbb{C}^{d \times d}\right)$. It is not difficult to prove that the condition (C1) in Theorem 6.1 is fulfilled if the function

$$t \mapsto \eta(t, \cdot), \quad t \in [s, s + h], \tag{6.25}$$

is Lipschitz continuous (recall that the space $NBV\left([-\tau, 0], \mathbb{C}^{d \times d}\right)$ is equipped with the total variation norm). For a nonautonomous DDE (2.8), i.e.,

$$x'(t) = A(t)x(t) + \sum_{k=1}^{p} B_k(t) x\,(t-\tau_k) + \sum_{k=1}^{p} \int_{-\tau_k}^{-\tau_{k-1}} C_k\,(t,\theta)\, x\,(t+\theta)\,\mathrm{d}\theta,$$

the function (6.25) is Lipschitz continuous if the functions A and $B_k, k = 1, \ldots, p$, are Lipschitz continuous in $[s, s+h]$ and, for $k = 1, \ldots, p$, there exist functions $m_k \in L^1([-\tau_k, -\tau_{k-1}], \mathbb{R})$ such that $\|C_k(t_1, \theta) - C_k(t_2, \theta)\| \le m_k(\theta) |t_1 - t_2|$ for all $t_1, t_2 \in [s, s+h]$ and for almost all $\theta \in [-\tau_k, -\tau_{k-1}]$.

On the other hand, similarly to what we have already seen in Chap. 5, the condition (C2) is satisfied if, for any index N, Ω_N^+ is a mesh of Chebyshev zero or type-I nodes, i.e.,

$$t_{N,i} = \frac{h}{2}\left(1 - \cos\left(\frac{(2n-1)\pi}{2N}\right)\right),\ i = 1, \ldots, N,$$

for which $\|L_N^+\| = O(\log N)$, $N \to \infty$. Such nodes are used in the implementation of the pseudospectral collocation method given in Part III, see Sect. 7.3.1.

By summarizing: the discussion given in this subsection shows that, under the conditions (C1) and (C2) of Theorem 6.1, the Eq. (6.10) has a unique solution Z_N^*, for any positive integer M, for any positive integer $N \ge N_0$, N_0 given in Theorem 6.1, and for any $\Phi \in X_M$. Thus, the finite dimensional operator $T_{M,N}$ given in (6.9) turns out to be well-defined.

6.3 Convergence Analysis

In this section, we study how the nonzero eigenvalues of the infinite dimensional operator $T : X \to X$, which are the important eigenvalues in stability studies (see Part I, Chap. 1), are approximated by the nonzero eigenvalues of the finite dimensional operator $T_{M,N} : X_M \to X_M$.

We assume that the conditions (C1) and (C2) of Theorem 6.1 are fulfilled and that $N \ge N_0$ holds, where N_0 is given in Theorem 6.1, so that $T_{M,N}$ is well-defined.

When we try to compare T and $T_{M,N}$, the main difficulty is that they are not defined on the same space. So, now we introduce another operator, denoted $\widehat{T}_{M,N}$, with the same nonzero eigenvalues of $T_{M,N}$, which is defined on the space X where T is defined. To this aim, we introduce the following lemma. Here, in this lemma, for *partial multiplicities* of an eigenvalue of finite type we mean the lengths of its Jordan chains ([73, 145]). Observe that the algebraic multiplicity is the sum of the partial multiplicities, whereas the geometric multiplicity is the number of Jordan chains.

Lemma 6.1 *Let Y and Z be linear spaces and let $A : Y \to Y$, $R : Z \to Y$, and $P : Y \to Z$ be linear operators. If*

$$RP = I_Y, \tag{6.26}$$

6.3 Convergence Analysis

then A has the same nonzero eigenvalues, with the same geometric and partial multiplicities, of

$$B = PAR : Z \to Z.$$

Moreover, if v is an eigenvector of A relevant to a nonzero eigenvalue λ, then Pv is an eigenvector of B relevant to the same eigenvalue λ.

Proof Assume (6.26) and let $\lambda \in \mathbb{C} \setminus \{0\}$. We begin by showing that

$$P(\ker(\lambda I_Y - A)) = \ker(\lambda I_Z - B). \tag{6.27}$$

We prove the inclusion $\subseteq$. Let $y \in Y$ such that $Ay = \lambda y$ and let $z = Py$. Then $Bz = PARz = PARPy = PAy = \lambda Py = \lambda z$. We prove the opposite inclusion $\supseteq$. Let $z \in Z$ such that $Bz = \lambda z$. Since $\lambda \neq 0$, we have $z = Py$, where $y = \frac{1}{\lambda} ARz$ and $\lambda y = ARz = ARPy = Ay$.

Now, we prove that

$$\dim \ker(\lambda I_Y - A) = \dim \ker(\lambda I_Z - B). \tag{6.28}$$

Let $y_1, \ldots, y_n$ be linearly independent elements of $\ker(\lambda I_Y - A)$. We have shown in (6.27) that $z_i = Py_i \in \ker(\lambda I_Z - B), i = 1, \ldots, n$. By (6.26), we obtain

$$\sum_{i=1}^{n} \alpha_i z_i = P \sum_{i=1}^{n} \alpha_i y_i = 0 \implies \sum_{i=1}^{n} \alpha_i y_i = 0 \implies \alpha_i = 0, \quad i = 1, \ldots, n,$$

and then the elements $z_1, \ldots, z_n$ are linearly independent. Vice-versa, let $z_1, \ldots, z_n$ be linearly independent elements of $\ker(\lambda I_Z - B)$. We have shown in (6.27) that $z_i = Py_i, i = 1, \ldots, n$, where $y_i \in \ker(\lambda I_Y - A)$. Since

$$\sum_{i=1}^{n} \alpha_i y_i = 0 \implies P \sum_{i=1}^{n} \alpha_i y_i = \sum_{i=1}^{n} \alpha_i z_i = 0 \implies \alpha_i = 0, \quad i = 1, \ldots, n,$$

the elements $y_1, \ldots, y_n$ are linearly independent. The relation (6.28) is thus proved.

By using (6.28), we can say that λ is an eigenvalue of A with geometric multiplicity g if and only if λ is an eigenvalue of B with geometric multiplicity g. By (6.27) and (6.26), we then obtain that if λ is an eigenvalue of A and v is a relevant eigenvector, then Pv is eigenvector of B relevant to λ (we have $Pv \neq 0$, otherwise $RPv = v = 0$).

It remains to prove that if λ is an eigenvalue of A, then it conserves the same partial multiplicities when considered as an eigenvalue of B. This is proved by showing that there is a one-to-one correspondence between Jordan chains of A and B relevant to λ.

Let $y_1, \ldots, y_n$ be a Jordan chain of A relevant to λ. Then $z_1 = Py_1, \ldots, z_n = Py_n$ is a Jordan chain of B relevant to λ. In fact, z_1 is an eigenvector of B and $(\lambda I_Z - B)z_{i+1} = (\lambda I_Z - B)Py_{i+1} = P(\lambda I_Y - A)y_{i+1} = z_i, i = 0, \ldots, n-1$.

Now, we show that the map

$$\begin{cases} \mathfrak{J} : \{\text{Jordan chain of } A \text{ relevant to } \lambda\} \to \{\text{Jordan chain of } B \text{ relevant to } \lambda\} \\ \mathfrak{J}(y_1, \ldots, y_n) = (Py_1, \ldots, Py_n) \end{cases}$$

is bijective. The injectivity follows by (6.26). Below, we prove that the map is surjective.

Let $z_1, \ldots, z_n$ be a Jordan chain of B relevant to λ. By (6.27), we have $z_1 = Py_1$ for some y_1 eigenvector of A. Note that, for $i = 0, \ldots, n-1$, if $z_i = Py_i$ for some $y_i \in Y$, then $z_{i+1} = Py_{i+1}$ for some $y_{i+1} \in Y$. In fact, $(\lambda I_Z - B)z_{i+1} = z_i = Py_i$ and so $z_{i+1} = P\left(\frac{1}{\lambda}(ARz_{i+1} + y_i)\right)$. We conclude that $z_i = Py_i, i = 1, \ldots, n$, for some $y_i \in Y, i = 1, \ldots, n$, and y_1 eigenvector of A. The sequence $y_1, \ldots, y_n$ is a Jordan chain for A. In fact, $Py_i = z_i = (\lambda I_Z - B)z_{i+1} = (\lambda I_Z - B)Py_{i+1} = P(\lambda I_Y - A)y_{i+1}, i = 0, \ldots, n-1$, and then, by (6.26), $y_i = (\lambda I_Y - A)y_{i+1}, i = 0, \ldots, n-1$. $\square$

By applying the previous proposition with $Y = X_M, Z = X, A = T_{M,N}, R = R_M$ and $P = P_M$, we obtain the following important result.

Theorem 6.2 *The finite dimensional operator $T_{M,N}$ has the same nonzero eigenvalues, with the same geometric and partial multiplicities, of the operator*

$$\widehat{T}_{M,N} = P_M T_{M,N} R_M : X \to X. \tag{6.29}$$

Moreover, if Φ is an eigenvector of $T_{M,N}$ relevant to a nonzero eigenvalue μ, then $P_M \Phi$ is an eigenfunction of $\widehat{T}_{M,N}$ relevant to μ.

The previous theorem says that in order to compare the nonzero eigenvalues of T with the nonzero eigenvalues of $T_{M,N}$, it is sufficient to compare the nonzero eigenvalues of T with the nonzero eigenvalues of $\widehat{T}_{M,N}$ defined in (6.29).

By using (6.9) and (6.10), we see that $\widehat{T}_{M,N}$ can be factorized as $\widehat{T}_{M,N} = \mathscr{L}_M \widehat{T}_N \mathscr{L}_M$, where $\widehat{T}_N : X \to X$ is given by

$$\widehat{T}_N \varphi = V(\varphi, w_N^*)_h, \quad \varphi \in X, \tag{6.30}$$

where $w_N^* \in X^+$ is the unique solution of (6.21) with $\phi = \varphi$. Such an equation has a unique solution since the conditions (C1) and (C2) of Theorem 6.1 have been assumed.

In the next section, we analyze the convergence of the nonzero eigenvalues of $\widehat{T}_N$ to the nonzero eigenvalues of T, as $N \to \infty$. Then, in the successive section, we study the convergence of the nonzero eigenvalues of $\widehat{T}_{M,N}$ to the nonzero eigenvalues of T, as $M, N \to \infty$.

6.3.1 Convergence of the Eigenvalues of $\widehat{T}_N$

Before presenting the convergence analysis of the nonzero eigenvalues of $\widehat{T}_N$, we establish a very useful result.

Lemma 6.2 *Let Y be a linear space, let $A : Y \to Y$ be a linear operator and let $\lambda \in \mathbb{C}$. If Z is a subspace of Y such that*

(1) $A(Z) \subseteq Z$;
(2) *for any* $y, z \in Y$, $(\lambda I_Y - A)y = z$, $z \in Z \implies y \in Z$;

then λ is an eigenvalue of geometric multiplicity g of A if and only if λ is an eigenvalue of $A|_Z : Z \to Z$ of geometric multiplicity g. Moreover, the eigenvectors relevant to the eigenvalue λ are the same for A and $A|_Z$ and also the partial multiplicities of λ are the same.

Proof Let Z be a subspace of Y satisfying the conditions (1) and (2). By (1), we have $A|_Z : Z \to Z$. Since $\ker(\lambda I_Y - A) \subseteq Z$, by (2) we obtain $\ker(\lambda I_Z - A|_Z) = \ker(\lambda I_Y - A)$. Hence, λ is an eigenvalue of A of geometric multiplicity $g(\lambda)$ if and only λ is an eigenvalue of $A|_Z$ of geometric multiplicity $g(\lambda)$. Moreover, the eigenvectors relevant to the eigenvalue λ are the same for A and $A|_Z$. Finally, the assertion concerning the partial multiplicities of the eigenvalue λ follows from the fact that $y_1, \ldots, y_n$ is a Jordan chain of A relevant to λ if and only if $y_1, \ldots, y_n$ is a Jordan chain of $A|_Z$ relevant to λ. This fact is a consequence of (2). □

Remark 6.1 Observe that, for $\lambda \neq 0$, the conditions (1) and (2) in the previous lemma are satisfied if the range $A(Y)$ of A is contained in Z.

Now, similarly to the subspace X^+_{Lip} in Sect. 6.2, we introduce the subspace X_{Lip} of X given by the Lipschitz continuous functions of X. The space X_{Lip} is equipped with the norm

$$\|\varphi\|_{X_{\text{Lip}}} = \text{Lip}(\varphi) + \|\varphi\|_X, \quad \varphi \in X_{\text{Lip}},$$

where $\text{Lip}(\varphi)$ denotes the Lipschitz constant of φ. With such a norm, X_{Lip} is a Banach space.

Observe that the map V in (6.1) has the following two properties:

$$V(\phi, z)_h \in X_{\text{Lip}}, \quad \phi \in X_{\text{Lip}}, \quad z \in X^+, \tag{6.31}$$

and

$$V(\phi, z)_h \in X_{\text{Lip}}, \quad \phi \in X, \quad z \in X^+, \quad h \geq \tau. \tag{6.32}$$

The first step in the analysis of the convergence of the nonzero eigenvalues of $\widehat{T}_N$ is to observe that we can apply Lemma 6.2 with $Y = X$, $A = T$, as well as $A = \widehat{T}_N$, $\lambda \neq 0$ and $Z = X_{\text{Lip}}$.

The condition (1) follows by (6.31). The condition (2) easily follows by (6.32) in the case $h \geq \tau$. In the case $h < \tau$, for $A = T$, one has to write the equation $(\lambda I_X - T)\varphi = \psi$, where $\varphi \in X$ and $\psi \in X_{\text{Lip}}$, as $V(\varphi, z^*)(h+\theta) + \psi(\theta) = \lambda\varphi(\theta)$, $\theta \in [-\tau, 0]$, i.e.,

$$\begin{cases} \varphi(0) + \displaystyle\int_0^{h+\theta} z^*(t)\,dt + \psi(\theta) = \lambda\varphi(\theta), & \theta \in [-h, 0], \\ \varphi(h+\theta) + \psi(\theta) = \lambda\varphi(\theta), & \theta \in [-\tau, -h]. \end{cases}$$

Now, it is not difficult to prove by induction that φ is Lipschitz continuous on each interval $[-(i+1)h, -ih]$, $i = 0, 1, 2, \ldots$ The same argument works for $A = \widehat{T}_N$.

Therefore, $T : X \to X$ and $\widehat{T}_N : X \to X$ have the same nonzero eigenvalues, with the same geometric and partial multiplicities, and the same relevant eigenvectors of the restrictions $T|_{X_{\text{Lip}}} : X_{\text{Lip}} \to X_{\text{Lip}}$ and $\widehat{T}_N|_{X_{\text{Lip}}} : X_{\text{Lip}} \to X_{\text{Lip}}$, respectively.

The second step in our analysis of the convergence is the following proposition.

Proposition 6.2 *If, in addition to the conditions (C1) and (C2) in Theorem 6.1, we also assume the condition*

(C3)   *the operator $\mathscr{F}_s V_1 : X \to X^+$ is such that $\mathscr{F}_s V_1(X_{\text{Lip}}) \subseteq X^+_{\text{Lip}}$ and $\mathscr{F}_s V_1|_{X_{\text{Lip}}} : X_{\text{Lip}} \to X^+_{\text{Lip}}$ is bounded;*

then

$$\|\widehat{T}_N|_{X_{\text{Lip}}} - T|_{X_{\text{Lip}}}\| \to 0, \quad N \to \infty.$$

Proof Let $\varphi \in X_{\text{Lip}}$. By recalling the form (6.3) of T and the definition (6.30) of $\widehat{T}_N$, we have $\widehat{T}_N\varphi - T\varphi = \left(V(\varphi, w_N^*) - V(\varphi, z^*)\right)_h = V_2(w_N^* - z^*)_h$, where w_N^* is the solution of $w^* = \mathscr{L}_N^+ \mathscr{F}_s V(\varphi, w^*)$ and z^* is the solution of

$$z^* = \mathscr{F}_s V(\varphi, z^*). \tag{6.33}$$

Thus,

$$\|\widehat{T}_N\varphi - T\varphi\|_{X_{\text{Lip}}} \leq (1+h)\|w_N^* - z^*\|_{X^+}. \tag{6.34}$$

The conditions (C1) and (C3) imply $z^* \in X^+_{\text{Lip}}$. By using the estimate (6.23), we obtain

$$\begin{aligned} \|w_N^* - z^*\|_{X^+} &\leq 2\left\|(I_{X^+} - \mathscr{F}_s V_2)^{-1}\right\| \left\|\mathscr{L}_N^+ z^* - z^*\right\|_{X^+} \\ &\leq 2\left\|(I_{X^+} - \mathscr{F}_s V_2)^{-1}\right\| \left\|(\mathscr{L}_N^+ - I_{X^+})|_{X^+_{\text{Lip}}}\right\|_{X^+ \leftarrow X^+_{\text{Lip}}} \|z^*\|_{X^+_{\text{Lip}}}. \end{aligned} \tag{6.35}$$

6.3 Convergence Analysis

Since $\varphi \in X_{\text{Lip}}$ and $z^* \in X_{\text{Lip}}^+$, we can rewrite (6.33) as $\left(I_{X_{\text{Lip}}^+} - \mathscr{F}_s V_2|_{X_{\text{Lip}}^+}\right) z^* = \mathscr{F}_s V_1|_{X_{\text{Lip}}} \varphi$. As we have observed at the beginning of the proof of Theorem 6.1, the operator $I_{X^+} - \mathscr{F}_s V_2$ is invertible. By (C1), we have that also the operator $I_{X_{\text{Lip}}^+} - \mathscr{F}_s V_2|_{X_{\text{Lip}}^+} : X_{\text{Lip}}^+ \to X_{\text{Lip}}^+$ is invertible. Thus $z^* = \left(I_{X_{\text{Lip}}^+} - \mathscr{F}_s V_2|_{X_{\text{Lip}}^+}\right)^{-1} \mathscr{F}_s V_1|_{X_{\text{Lip}}} \varphi$ and then

$$\|z^*\|_{X_{\text{Lip}}^+} \leq \left\|(I_{X_{\text{Lip}}^+} - \mathscr{F}_s V_2|_{X_{\text{Lip}}^+})^{-1}\right\| \left\|\mathscr{F}_s V_1|_{X_{\text{Lip}}}\right\|_{X_{\text{Lip}}^+ \leftarrow X_{\text{Lip}}} \|\varphi\|_{X_{\text{Lip}}}. \quad (6.36)$$

By using (C2) and the interpolation theory result (6.24) given in Theorem 6.1, we obtain

$$\left\|(\mathscr{L}_N^+ - I_{X^+})|_{X_{\text{Lip}}^+}\right\|_{X^+ \leftarrow X_{\text{Lip}}^+} \to 0, \quad N \to \infty.$$

By using this fact, the thesis immediately follows by (6.34), (6.35) and (6.36). □

Similarly to the condition (C1) of Theorem 6.1, the condition (C3) of Proposition 6.2 is fulfilled if the function (6.25) is Lipschitz continuous.

The final step of our analysis is to present a standard result on the approximations of eigenvalues of infinite dimensional linear and bounded operators given in [60, Theorem 6.7].

Theorem 6.3 *Let A be a linear and bounded operator on the Banach space Y and let $\{A_N\}$ be a sequence of linear and bounded operators on Y such that*

$$\|A_N - A\| \to 0, \quad N \to \infty. \quad (6.37)$$

If $\lambda \in \mathbb{C}$ is an isolated eigenvalue of A with finite algebraic multiplicity ν and ascent l, and Δ is a neighborhood of λ such that λ is the unique eigenvalue of A in Δ, then there exists an index $\overline{N}$ such that, for any index $N \geq \overline{N}$, A_N has in Δ exactly ν eigenvalues $\lambda_{N,j}$, $j = 1, \ldots, \nu$, counting their multiplicities. Moreover, by setting

$$\varepsilon_N := \left\|(A_N - A)|_{\mathscr{E}_\lambda}\right\|_{Y \leftarrow \mathscr{E}_\lambda}, \quad (6.38)$$

where $\mathscr{E}_\lambda$ is the generalized eigenspace of λ equipped with the norm $\|\cdot\|_Y$ restricted to $\mathscr{E}_\lambda$, we have

$$\max_{j=1,\ldots,\nu} |\lambda_{N,j} - \lambda| = O\left(\varepsilon_N^{1/l}\right), \quad N \to \infty.$$

Indeed, in [60, Theorem 6.7], this results holds under a condition on the approximating sequence $\{A_N\}$ weaker than the norm convergence (6.37) and known as *strong stable convergence*. Moreover, also the convergence of the eigenvectors is considered there. Theorem 6.3 says that the order of convergence to zero as $N \to \infty$ of the eigenvalues depends on the order of convergence to zero of ε_N, namely the norm of the error $A_N - A$ as restricted to the generalized eigenspace $\mathcal{E}_\lambda$.

Eventually, we are now ready to give the convergence result of the nonzero eigenvalues of $\widehat{T}_N$ to the nonzero eigenvalues of T, as $N \to \infty$.

Theorem 6.4 *Assume the conditions (C1) and (C2) of Theorem 6.1 and the condition (C3) of Proposition 6.2. Let $\mu^* \in \mathbb{C} \setminus \{0\}$ be an eigenvalue of T with finite algebraic multiplicity ν^* and ascent l^*, and let Δ be a neighborhood of μ^* such that μ^* is the unique eigenvalue of T in Δ. Then, there exists a positive integer N_1 with $N_1 \geq N_0$, where N_0 is given in Theorem 6.1, such that, for any index $N \geq N_1$, $\widehat{T}_N$ has in Δ exactly ν^* eigenvalues $\mu^*_{N,j}$, $j = 1, \ldots, \nu^*$, counting their multiplicities. Moreover, if*

(C4) *for any $\varphi \in \mathcal{E}_{\mu^*}$, where $\mathcal{E}_{\mu^*}$ is the generalized eigenspace of T relevant to μ^*, the solution z^* of (6.4), i.e.,*

$$z^* = \mathscr{F}_s V(\varphi, z^*) \tag{6.39}$$

is of class C^p;

then

$$\max_{j=1,\ldots,\nu^*} |\mu_{N,j}{}^* - \mu^*| = o\left(\frac{1}{N^{\frac{p-1}{l^*}}}\right), \quad N \to \infty. \tag{6.40}$$

Proof By recalling that T and $\widehat{T}_N$ have the same nonzero eigenvalues, with the same geometric and partial multiplicities of the restrictions $T|_{X_{\text{Lip}}}$ and $\widehat{T}_N|_{X_{\text{Lip}}}$, respectively, we apply Theorem 6.3 with $Y = X_{\text{Lip}}$, $A = T|_{X_{\text{Lip}}}$ and $A_N = \widehat{T}_N|_{X_{\text{Lip}}}$.

Under the assumptions (C1), (C2) and (C3), the condition (6.37) holds by Proposition 6.2.

Now, we analyze the error ε_N given in (6.38). Let $\psi_1^*, \ldots, \psi_{\nu^*}{}^*$ be a basis for $\mathcal{E}_{\mu^*}$. By writing an arbitrary element $\varphi \in \mathcal{E}_{\mu^*}$ as $\varphi = \sum_{i=1}^{\nu^*} \alpha_i(\varphi) \psi_i^*$, where $\alpha_i(\varphi) \in \mathbb{C}$, $i = 1, \ldots, \nu^*$, we have

$$\left\| \widehat{T}_N \varphi - T\varphi \right\|_{X_{\text{Lip}}} \leq \max_{i=1,\ldots,\nu^*} |\alpha_i(\varphi)| \sum_{i=1}^{\nu^*} \left\| \widehat{T}_N \psi_i^* - T\psi^*_i \right\|_{X_{\text{Lip}}}.$$

Since $\varphi \mapsto \max_{i=1,\ldots,\nu^*} |\alpha_i(\varphi)|$, $\varphi \in \mathcal{E}_{\mu^*}$, is a norm on $\mathcal{E}_{\mu^*}$, it is equivalent to the norm $\|\cdot\|_{X_{\text{Lip}}}$ restricted to $\mathcal{E}_{\mu^*}$. Thus, there exists a constant C such that $\max_{i=1,\ldots,\nu^*} |\alpha_i(\varphi)| \leq C \|\varphi\|_{X_{\text{Lip}}}$ for any $\varphi \in \mathcal{E}_{\mu^*}$. Therefore, we obtain

$$\varepsilon_N = \left\| (\widehat{T}_N - T)|_{X_{\text{Lip}}} \right\|_{X_{\text{Lip}} \leftarrow \mathcal{E}_{\mu^*}} \leq C \sum_{i=1}^{\nu^*} \left\| \widehat{T}_N \psi_i^* - T\psi^*_i \right\|_{X_{\text{Lip}}}. \tag{6.41}$$

6.3 Convergence Analysis

Let $i = 1, \ldots, \nu^*$. As in (6.34) and (6.35), we obtain

$$\left\| \widehat{T}_N \psi_i^* - T\psi_i^* \right\|_{X_{\text{Lip}}} \leq 2(1+h) \left\| (I_{X^+} - \mathscr{F}_s V_2)^{-1} \right\| \cdot \left\| \mathscr{L}_N^+ z_i^* - z_i^* \right\|_{X^+}, \quad (6.42)$$

where z_i^* is the solution of $z_i^* = V(\psi_i^*, z_i^*)$. Since z_i^* is of class C^p by condition (C4), interpolation theory (see [163, Theorem 1.5]) provide us with the bound

$$\left\| \mathscr{L}_N^+ z_i^* - z_i^* \right\|_{X^+} \leq (1 + \|\mathscr{L}_N^+\|) \left(\frac{h}{2}\right)^p \cdot \frac{c_p}{(N-1)^p} \left\| (z_i^*)^{(p)} \right\|_{X_{\text{Lip}}},$$

where c_p depends only on p. Then, the condition (C2) implies

$$\left\| \mathscr{L}_N^+ z_i^* - z_i^* \right\|_{X^+} = o\left(\frac{1}{N^{p-1}}\right), \quad N \to \infty. \quad (6.43)$$

By (6.41), (6.42) and (6.43) we can conclude that

$$\varepsilon_N = o\left(\frac{1}{N^{p-1}}\right), \quad N \to \infty, \quad (6.44)$$

and then (6.40) in the thesis follow by the estimate given in Theorem 6.3. □

6.3.2 Convergence of the Eigenvalues of $\widehat{T}_{M,N}$

After the study accomplished in the previous section regarding the convergence of the nonzero eigenvalues of $\widehat{T}_N$ as $N \to \infty$, we can consider now the convergence of the nonzero eigenvalues of $T_{M,N}$, as $M, N \to \infty$.

We begin by comparing the nonzero eigenvalues, as well as the relevant eigenvectors, of $\widehat{T}_{M,N}$ and $\widehat{T}_N$.

Theorem 6.5 *Let M and N be indices of discretization with $N \geq N_0$, N_0 given in Theorem 6.1. If $M \geq N$, then $\widehat{T}_{M,N}$ has the same nonzero eigenvalues, with the same geometric and partial multiplicities, and the same relevant eigenvectors of $\widehat{T}_N$.*

Proof Assume $M \geq N$. First, we consider the case $h \geq \tau$. We denote by Π_k and Π_k^+, where k is a nonnegative integer, the subspaces of X and X^+, respectively, of the k-degree $\mathbb{C}^d$-valued polynomial functions. Note that the solution w_N^* of (6.21) belongs to Π_{N-1}^+. Then, since $h \geq \tau$, we have that, for any $\varphi \in X$, $\widehat{T}_N \varphi = V(\varphi, w_N^*)_h \in \Pi_N$. Since $\widehat{T}_N$ has range contained in Π_N and $M \geq N$, both $\widehat{T}_{M,N} = L_M \widehat{T}_N L_M$ and $\widehat{T}_N$ have range contained in Π_M. Therefore, we can apply Lemma 6.2 with $Y = X$, $A = \widehat{T}_{M,N}$, as well as $A = \widehat{T}_N$, and $Z = \Pi_M$, by using Remark 6.1. We conclude that $\widehat{T}_{M,N}$ and $\widehat{T}_N$ have the same nonzero eigenvalues, with the same geometric and partial multiplicities, and the same relevant eigenvectors of the restrictions $\widehat{T}_{M,N}|_{\Pi_M}$:

$\Pi_M \to \Pi_M$ and $\widehat{T}_N|_{\Pi_M} : \Pi_M \to \Pi_M$, respectively. The thesis follows from the fact that $\widehat{T}_{M,N}|_{\Pi_M} = \mathscr{L}_M \widehat{T}_N \mathscr{L}_M|_{\Pi_M} = \widehat{T}_N|_{\Pi_M}$.

Now, we consider the other case $h < \tau$. We denote by Π_k^{pw} the subspace of X of the functions that are piecewise k-degree $\mathbb{C}^d$-valued polynomials on the intervals $[\theta^{(q+1)}, \theta^{(q)}]$, $q = 0, \ldots, Q - 1$. We can apply Lemma 6.2, with $Y = X$, $A = \widehat{T}_N$ and $Z = \Pi_M^{\mathrm{pw}}$. Since $M \geq N$, the condition 1) is fulfilled. By decomposing the equation $(\lambda I_X - \widehat{T}_N)\varphi = \psi$, where $\varphi \in X$ and $\psi \in \Pi_M^{\mathrm{pw}}$, as

$$\begin{cases} \varphi(0) + \int_0^{h+\theta} w_N^*(t)\mathrm{d}t + \psi(\theta) = \lambda\varphi(\theta), & \theta \in [-h, 0], \\ \varphi(h + \theta) + \psi(\theta) = \lambda\varphi(\theta), & \theta \in [-\tau, -h], \end{cases}$$

we see that the condition (2) holds whenever $\lambda \neq 0$. Hence, $\widehat{T}_N : X \to X$ has the same nonzero eigenvalues, with the same geometric and partial multiplicities, and the same relevant eigenvectors of the restriction $\widehat{T}_N|_{\Pi_M^{\mathrm{pw}}} : \Pi_M^{\mathrm{pw}} \to \Pi_M^{\mathrm{pw}}$. Moreover, we can apply Lemma 6.2 with $A = \widehat{T}_{M,N}$, instead of $A = \widehat{T}_N$. In fact, observe that the range of $\widehat{T}_{M,N} = \mathscr{L}_M \widehat{T}_N \mathscr{L}_M$ is contained in Π_M^{pw} and recall Remark 6.1. Thus, $\widehat{T}_{M,N} : X \to X$ has the same nonzero eigenvalues, with the same geometric and partial multiplicities, and the same relevant eigenvectors of the restriction $\widehat{T}_{M,N}|_{\Pi_M^{\mathrm{pw}}} : \Pi_M^{\mathrm{pw}} \to \Pi_M^{\mathrm{pw}}$. Now, similarly to the case $h \geq \tau$, the thesis follows from the fact that $\widehat{T}_{M,N}|_{\Pi_M^{\mathrm{pw}}} = \mathscr{L}_M \widehat{T}_N \mathscr{L}_M|_{\Pi_M^{\mathrm{pw}}} = \widehat{T}_N|_{\Pi_M^{\mathrm{pw}}}$. □

Thus, by using the previous Theorems 6.2, 6.4 and 6.5, we obtain our final convergence theorem.

Theorem 6.6 *Assume the conditions (C1) and (C2) of Theorem 6.1 and the condition (C3) of Proposition 6.2. Let $\mu^* \in \mathbb{C} \setminus \{0\}$ be an eigenvalue of T with finite algebraic multiplicity ν^* and ascent l^*, and let Δ be a neighborhood of μ^* such that μ^* is the unique eigenvalue of T in Δ. Then there exists a positive integer N_1 with $N_1 \geq N_0$, where N_0 is given in Theorem 6.1, such that, for any index $N \geq N_1$ and for any index $M \geq N$, $T_{M,N}$ has in Δ exactly ν^* eigenvalues $\mu_{M,N,j}^*$, $j = 1, \ldots, \nu^*$, counting their multiplicities. Moreover, if (C4) holds, then*

$$\max_{j=1,\ldots,\nu^*} |\mu_{M,N,j}^* - \mu^*| = o\left(\frac{1}{N^{\frac{p-1}{l^*}}}\right), \quad N \to \infty, \; M \geq N. \tag{6.45}$$

The following is a corollary of the previous result.

Theorem 6.7 *Assume the conditions (C1) and (C2) of Theorem 6.1 and the condition (C3) of Proposition 6.2. Let B be a bounded region of $\mathbb{C}$, whose closure does not contain 0, and let $\mu_1^*, \ldots, \mu_K^*$ be eigenvalues of T in B of multiplicity $\nu_1^*, \ldots, \nu_K^*$ and ascent $l_1^*, \ldots, l_K^*$, respectively. Then, there exists an index N_2 with $N_2 \geq N_0$, where N_0 is given in Theorem 6.1, such that, for any indices $N \geq N_2$ and $M \geq N$ and*

6.3 Convergence Analysis

for each $k = 1, \ldots, K$, $T_{M,N}$ has eigenvalues $\mu^*_{M,N,k,j}$, $j = 1, \ldots, \nu^*_k$, counting their multiplicities. These are all the eigenvalues of $T_{M,N}$ in B. Moreover, if

(C5) for any $k = 1, \ldots, K$ and for any $\varphi \in \mathcal{E}_{\mu^*_k}$, the solution z^* of (6.4) is of class C^p;

then

$$\max_{\substack{k=1,\ldots,K \\ j=1,\ldots,\nu^*_k}} \left|\mu^*_{M,N,k,j} - \mu^*\right| = o\left(\frac{1}{N^{\frac{p-1}{l^*}}}\right), \quad N \to \infty, \ M \geq N. \tag{6.46}$$

Proof Partition B in k subset $\Delta_1, \ldots, \Delta_k$ so that, for any $k = 1, \ldots, k$, μ^*_k is the unique eigenvalue of T in Δ_k. For any $k = 1, \ldots, k$, let $N_{1,k}$ be the index N_1 of Theorem 6.6 with $\mu^* = \mu^*_k$ and $\Delta = \Delta_k$. We take the maximum of the indices $N_{1,k}$, $k = 1, \ldots, K$, as index N_2. □

As observed in Part I, the asymptotic stability of the zero solution of a linear autonomous DDE, the asymptotic stability of the zero solution of a linear periodic DDE and the asymptotic stability of a periodic solution of a nonlinear autonomous DDE can be determined by the position of the nonzero eigenvalues of an evolution operator. In the first case, the evolution operator is $T(h, 0)$, $h > 0$ arbitrary, and in the second and third cases, the evolution operator is the monodromy operator $T(\omega, 0)$, where ω is the period of the linear periodic DDE or the period of the periodic solution of the nonlinear autonomous DDE.

It is easy to check that, in all the previous situations, the solution z^* of (6.39) is of class C^∞. Therefore, the pseudospectral collocation method of the SO approach exhibits an infinite order of convergence for the approximated eigenvalues.

We also observe that in our convergence analysis there are no requirements about the nodes of the mesh Ω_M on $[-\tau, 0]$, except for $M \geq N$. Anyway, the mesh Ω_M that is be used in Part III is a mesh of Chebyshev nodes.

6.3.3 Quadrature for Distributed Delays

In case of a distributed delay, where the functional L has to be replaced with a functional $L_{M,N}$, expression of the use of a quadrature rule, we can repeat the previous analysis and conclude that in (6.44) a further term

$$o\left(\|\mathcal{L}^+_N\| \cdot \max_{i=1,\ldots,\nu^*} \|z^*_{M,N,i} - z^*\|_{X^+}\right)$$

has to be added, where $z^*_{M,N,i}$, $i = 1, \ldots, \nu^*$, is the solution of $z^* = \mathcal{F}_{s,M,N} V(\psi_i, z^*)$. Here, $\mathcal{F}_{s,M,N}$ is obtained by replacing L with $L_{M,N}$ in (6.2) and $\psi_1, \ldots, \psi_{\nu^*}$ is a basis for $\mathcal{E}(\mu^*)$.

So, in (6.45) we have a further error term

$$O\left(\left(\|\mathscr{L}_N^+\| \cdot \max_{i=1,\ldots,\nu^*} \|z_{M,N,i}^* - z^*\|_{X^+}\right)^{1/\ell}\right)$$

and in (6.46) a further error term

$$O\left(\left(\|\mathscr{L}_N^+\| \cdot \max_{i=1,\ldots,\nu^*} \|z_{M,N,i}^* - z^*\|_{X^+}\right)^{1/\ell^*}\right).$$

6.4 Other Methods

In the literature, one can find many methods of the SO approach, where an evolution operator is discretized in a finite dimensional operator. However, most of them discretizes an evolution operator $T(s+h,s)$ with $h = \tau$. This particular situation makes things easier, since, in practice, the introduction of the space X^+ is no longer necessary. It is clear that any method for integrating differential equations can be used as a method of the SO approach. So, in the autonomous case, the solution operator is discretized by linear multistep methods in [80] and by Runge-Kutta methods in [31]. Methods for the periodic case can be found in [51–54, 136].

Part III
Implementation and Applications

The numerical methods presented in Part II for linear DDEs (2.4) are adapted in this last part of the book to system (2.8). The latter describes with sufficient generality the class of DDEs of interest in most applications, with a structure that is the most general and suitable to be treated numerically.

Chapter 7 of this part explains in detail how the methods are implemented in MATLAB. System (2.8) is taken here as a prototype, given its generality from the numerical point of view.

Nevertheless, even more freedom is left to the user if a slightly modified formulation is considered, especially when dealing with models having varying or uncertain parameters. Chapter 8 is based on this user's point of view. There we explain how to use the MATLAB codes for analyzing a benchmark set of case studies, as well as real-life applications.

One formulation or the other, the aim of this part is in guiding the interested reader to the understanding and application of the proposed algorithms.

Eventually, let us emphasize once more that models coming from applications can be either originally linear (autonomous or periodic) or can be obtained by linearizing nonlinear systems at specific solutions (equilibria or periodic). As a consequence, the analysis of (2.8) through the computation of the characteristic roots (autonomous case) or of the characteristic multipliers (periodic case) provide information, respectively, on the global stability of the zero solution for the linear problem itself or on the local stability of a specific solution for nonlinear ones linearized at the latter.

Part III
Implementation and Applications

Chapter 7
MATLAB Implementation

The book is provided with the following three MATLAB codes:

- myDDE.m;
- eigAM.m;
- eigTMN.m;

freely available [48]. The first one is a MATLAB m-file *script* that contains all the information necessary to define the linear DDE describing the model to be analyzed: it is the argument of Sect. 7.1. The second one is a MATLAB m-file *function* that implements the IG approach according to the (piecewise) pseudospectral differentiation method presented in Chap. 5: it is the argument of Sect. 7.2. The latter one is a MATLAB m-file *function* that implements the SO approach according to the pseudospectral collocation method presented in Chap. 6: it is the argument of Sect. 7.3.

The idea behind the division of the computational framework between a script (myDDE.m) and a function (either eigAM.m or eigTMN.m depending on the problem at hand, i.e., autonomous or nonautonomous) is to separate the implementation of the model from the implementation of the numerical methods. Indeed, the latter are contained exclusively in the functions, while the user is only required to fill the necessary fields in the script for defining the analyzed model. This way, also users not necessarily familiar with numerical analysis can use the codes for their experiments in a friendly fashion.

7.1 Introducing the Model in MATLAB

Throughout the book, the prototype model (2.8), i.e.,

$$x'(t) = A(t)x(t) + \sum_{k=1}^{p} B_k(t)x(t-\tau_k) + \sum_{k=1}^{p} \int_{-\tau_k}^{-\tau_{k-1}} C_k(t,\theta)x(t+\theta)\,d\theta \quad (7.1)$$

has been thought as the most general and suitable to describe and implement the proposed numerical approaches. This model, beyond the regularity of the coefficients required to guarantee the convergence of the approximated eigenvalues (see Chaps. 5 and 6), poses two constraints: first, all the coefficients and the delays are given and fixed and, second, all the delays (and the relevant coefficients) must be ordered according to $0 =: \tau_0 < \tau_1 < \cdots < \tau_p := \tau$. While optimal from the numerical point of view, this structure is restrictive from the applications point of view. In fact, coefficients and delays in real-life models may depend on varying or uncertain parameters, for the sake, e.g., of robust and bifurcation analysis. Moreover, the variation of these parameters can cause an exchange in the ordering of the delays.

As to clarify, let us briefly consider the DDE $x'(t) = ax(t) + b_1 x(t - \tau_1) + b_2 x(t - \tau_2)$, where a, b_1, and b_2 are fixed numbers while τ_1 and τ_2 are varying parameters. Being varying, it can happen that from an initial situation where $\tau_1 < \tau_2$, at some point it occurs that $\tau_2 < \tau_1$. Therefore, also the terms $b_1 x(t - \tau_1)$ and $b_2 x(t - \tau_2)$ shall be exchanged to adhere to the requirements of (7.1). Note that it can also occur that $\tau := \tau_1 = \tau_2 > 0$, reducing the DDE to $x'(t) = ax(t) + (b_1 + b_2)x(t - \tau)$, or even that $\tau_1 = \tau_2 = 0$, reducing to the ODE $x'(t) = (a + b_1 + b_2)x(t)$, whose only characteristic root is clearly $a + b_1 + b_2$ and no approximation at all is needed. A similar case from a real-life application is treated in Sect. 8.3, where a stability chart w.r.t. two varying discrete delays is presented. Eventually, it is not difficult to think about the wide panorama of possible situations if also distributed delay terms were considered (e.g., an integral vanishes when the integration extrema coincide).

In order to allow the treatment of DDEs with varying parameters, the script myDDE.m refers to the more general model

$$x'(t) = \tilde{A}(t)x(t) + \sum_{u=1}^{q} \tilde{B}_u(t)x(t - d_u) + \sum_{v=1}^{w} \int_{-l_v}^{-r_v} \tilde{C}_v(t, \theta)x(t + \theta)\,d\theta, \quad (7.2)$$

where all the coefficients and the delays can depend on a vector of parameters par, no relation is imposed between the discrete delays and the integration extrema of the distributed delay terms and, finally, no ordering is required among all the delays and extrema. The only (legitimate) constraint is that such delays and extrema must be nonnegative, otherwise the model would enter the class of advanced-retarded differential equations [11, 96, 140, 141, 166, 167], whose stability issues and relevant numerical treatment are different [6, 13, 40, 94] and beyond the target of this book.

Once the script myDDE.m is created according to (7.2), the user has just to give its name in input either to the function eigAM.m or to the function eigTMN.m, together with a vector par of the values of the possible parameters of the model. Other inputs to eigAM.m or eigTMN.m are described in the relevant sections. When both functions are executed, the first instruction loads the content of myDDE.m in the relevant workspace, making available all the necessary model data and parameters, and automatically converts model (7.2) into the structure of model (7.1). This procedure avoids in general repeated calls to external functions defining the model, thus

7.1 Introducing the Model in MATLAB

notably reducing the overall computational time and, above all, leaves a complete freedom of modeling to the user, who has not to worry about constraints or ordering schemes. The forthcoming Sects. 7.2 and 7.3 refer to the implementation of the numerical approaches according to (7.1), hence from the numerical point of view. Chapter 8, instead, presents a series of tests and applications whose relevant DDEs are described according to (7.2), hence from the user's point of view.

Now, let us analyze the script myDDE.m, so to guide the user into its compilation. The following is the content of the template (excluding the principal comment):

```
%% MEMO LIST OF POSSIBLE PARAMETERS
%par(1)=1st parameter;
%par(2)=2nd parameter;
%...

%% DIMENSION OF THE DDE
d=[]; %INPUT

%% CURRENT TIME TERM
Atilde=@(t,d,par) []; %INPUT: dxd matrix or
                     %call to Atilde.m

%% DISCRETE DELAY TERMS
dd=[]; %INPUT discrete delays row vector
       %dd=[d_{1},...,d_{q}]>=0
Btilde{1}=@(t,d,par) []; %INPUT: dxd matrix or
                        %call to Btilde1.m
%...
Btilde{q}=@(t,d,par) []; %INPUT: dxd matrix or
                        %call to Btildeq.m

%% DISTRIBUTED DELAY TERMS
l=[]; %INPUT left integration extrema row vector
      %l=[l_{1},....,l_{w}]>=0
r=[]; %INPUT right integration extrema row vector
      %r=[r_{1},....,r_{w}]>=0
Ctilde{1}=@(t,theta,d,par) []; %INPUT: dxd matrix or
                              %call to Ctilde1.m
%...
Ctilde{w}=@(t,theta,d,par) []; %INPUT: dxd matrix or
                              %call to Ctildew.m
```

Skipping the main comment (lines 1–3 in the m-file), which can be read by asking for help myDDE at the command prompt, a first commented section (lines 33–36) reports the list of possible parameters to be used as variable inputs in the sequel: it is just for convenience of recalling them. Note, in fact, that the vector par is not created (although used) in the script, but it will be available when the script will be loaded in either eigAM.m or eigTMN.m since par is given in input there.

The first true section of the script (lines 38–39) requires the input of the (fixed) positive integer d defining the number of equations in (7.2).

A second section (lines 41–42) concerns the input of the current time coefficient $\tilde{A}$. It is defined via a MATLAB *anonymous function* through the use of @, [1]. The main argument is the time t. Other arguments are the dimension d, so one can define, e.g., a null matrix through zeros(d), and possible parameters through the vector par.

A third section (lines 44–48) concerns the input of the discrete delay terms. The user is asked to create a row vector dd containing the nonnegative discrete delays $d_1, \ldots, d_q$ (line 45). These delays can be defined through the vector par and no ordering is required. The relevant coefficients $\tilde{B}_u$, $u = 1, \ldots, q$, are introduced via anonymous functions as for $\tilde{A}$ above. All these coefficients are elements of a MATLAB *cell-array* B, [3]: this simplifies the possible reordering in the conversion to (7.1).

Eventually, a last section (lines 50–55) concerns the input of the distributed delay terms. The user is required to create two row vectors l and r containing, respectively, the nonnegative "left" and "right" integration extrema $l_1, \ldots, l_w$ and $r_1, \ldots, r_w$ (lines 51–52). They can depend on the vector par and it can also be $l_v \le r_v$ for some or any $v = 1, \ldots, w$. Then, the relevant kernels $\tilde{C}_v$, $v = 1, \ldots, w$, are introduced via anonymous functions and cell arrays again, the only difference being the presence of the additional integration variable θ.

Consider the script myDDE_SISC_54.m, relevant to [38, Example 5.4]:

$$\begin{cases} x_1'(t) = -3x_1(t) + x_2(t) + x_1(t - d_1) \\ \quad + \int_{-0.3}^{-0.1} [2.25x_1(t+\theta) + 2.5x_2(t+\theta)]\,d\theta - \int_{-1}^{-0.5} x_1(t+\theta)\,d\theta, \\ x_2'(t) = -24.646x_1(t) - 35.430x_2(t) + 2.35553x_1(t-d_1) + 2.00365x_2(t-d_1) \\ \quad - \int_{-0.3}^{-0.1} c\theta^2 x_2(t+\theta)\,d\theta - \int_{-1}^{-0.5} x_2(t+\theta)\,d\theta, \end{cases}$$

as an example, here slightly modified to allow for varying parameters (the discrete delay d_1 and the coefficient c in the first distributed delay term of the second equation) and nonconstant kernels. The main content of the script is:

```
%% MEMO LIST OF POSSIBLE PARAMETERS
%par(1)=d_1;
%par(2)=c;

%% DIMENSION OF THE DDE
d=2;

%% CURRENT TIME TERM
Atilde=@(t,d,par) [-3,1;-24.646,-35.430];

%% DISCRETE DELAY TERMS
dd=par(1);
Btilde{1}=@(t,d,par) [1,0;2.35553,2.00365];
```

7.1 Introducing the Model in MATLAB

```
%% DISTRIBUTED DELAY TERMS
l=[.3,1];
r=[.1,.5];
Ctilde{1}=@(t,theta,d,par) [2.25,2.5;...
            0,-par(2)*theta.^2];
Ctilde{2}=@(t,theta,d,par) [-1,0;0,-1];
```

Further examples are given in Chap. 8 where, indeed, all the models proposed are accompanied with the description of the relevant script.

Some notes follow. First, observe that the script myDDE.m is the same independently of its use with eigAM.m or eigTMN.m. As the former is suitable only for autonomous problems, all the functions are implicitly intended as independent of time (as for the example above). Second, the user can define the coefficients of the model as external MATLAB functions. This may be necessary when the coefficients themselves come from previous computations. If so, the user has to build an external function, e.g., Btilde1.m, and write in the relevant line of myDDE.m

```
Btilde{1}=@(t,d,par) Btilde1(t,d,par);
```

Remark 7.1 The definitions of all the functions $\tilde{A}$, $\tilde{B}_u$, $u = 1, \ldots, q$, and $\tilde{C}_v$, $v = 1, \ldots, w$, in myDDE.m or externally have to be amenable of MATLAB element-wise evaluation, [2]. E.g., $\tilde{A}(t) = t^2$ has to be implemented as t.^2 and not t^2. See, in fact, Ctilde1 in the example above.

7.2 The Infinitesimal Generator Approach

As explained in Sect. 7.1, the implementation of the IG approach with pseudospectral differentiation methods developed in Chap. 5 is referred to the prototype model (7.1) for the autonomous case, i.e., with coefficients independent of the time t. The latter is obtained from (2.7) by choosing

$$L\varphi = A\varphi(0) + \sum_{k=1}^{p} B_k \varphi(-\tau_k) + \sum_{k=1}^{p} \int_{-\tau_k}^{-\tau_{k-1}} C_k(\theta)\varphi(\theta)\,d\theta, \quad \varphi \in X. \quad (7.3)$$

Let us remark that (7.1) can describe either an originally linear model or can be the result of the linearization of a nonlinear model around an equilibrium. Then the approximation of the characteristic roots gives information, respectively, on the global stability of the zero solution or on the local stability of the equilibrium (by the Principle of Linearized Stability, Theorem 3.5).

For the general linear functional (7.3) and according to Sect. 5.1, the matrix discretizing the infinitesimal generator reads

$$\mathscr{A}_M = \begin{pmatrix} a_0 & a_1 & \cdots & a_M \\ d_{1,0} & d_{1,1} & \cdots & d_{1,M} \\ \vdots & \vdots & \ddots & \vdots \\ d_{M,0} & d_{M,1} & \cdots & d_{M,M} \end{pmatrix} \in \mathbb{R}^{d(M+1) \times d(M+1)}, \tag{7.4}$$

where the discretization index M is a positive integer, $a_j = L(\ell_{M,j}(\cdot)I_d)$, $j = 0, 1, \ldots, M$, and $d_{m,j} = \ell'_{M,j}(\theta_{M,m})I_d$, $m = 1, \ldots, M$, $j = 0, 1, \ldots, M$, with $\ell_{M,0}, \ell_{M,1}, \ldots, \ell_{M,M}$ the Lagrange coefficients relevant to the mesh Ω_M in (5.1) discretizing the delay interval $[-\tau, 0]$, τ being the maximum delay.

Although it may not be evident yet, even though anticipated in Chap. 5, this discretization is suitable for DDEs with a single delay, i.e., $p = 1$ in (7.3). When $p > 1$, a piecewise approach as described in Sect. 5.2 is more efficient. Therefore, we proceed by illustrating, first, the implementation for the single discrete delay case in Sect. 7.2.1, second, the implementation for the single distributed delay case in Sect. 7.2.2 and, eventually, the piecewise implementation for the general case (7.3) in Sect. 7.2.3. The MATLAB code eigAM.m refers to the piecewise strategy and the following sections describe how to construct the matrix $\mathscr{A}_M$ step-by-step.

It is useful to remark that, according to the convergence analysis in Sect. 5.3 (see the end of Sect. 5.4), all the meshes implemented in eigAM.m are based on extremal Chebyshev nodes or, briefly, Chebyshev II nodes. Beyond satisfying (5.11) and thus guaranteeing the spectral accuracy (5.32), they exhibit the Lebesgue constant with the slowest increase w.r.t. their number (see, e.g., [186, Chap. 15]), a property with many positive consequences on interpolation and quadrature [184].

The overall structure of eigAM.m is organized as follows:

- the main comment (lines 2–22);
- the conversion of the model (7.2) as defined in myDDE.m into the DDE (7.1) (lines 24–27);
- the definition of the mesh Ω_M (lines 29–36);
- the construction of the matrix $\mathscr{A}_M$ (lines 38–66);
- the computation of the eigenvalues λ of $\mathscr{A}_M$ by eig, ordered by decreasing real part (lines 68–71);
- a set of auxiliary functions, see later on (lines 73–264).

The standard call is

```
[lambda,M]=eigA('myDDE',par,M);
```

where the inputs are the name of the script myDDE.m, a vector par of possible parameter values as explained in Sect. 7.1 and the positive integer M defining the discretization index. The outputs are the vector containing the $d(M+1)$ eigenvalues of $\mathscr{A}_M$ and the number M, with the meaning that $M+1$ is the total number of nodes actually used. In fact, the integer M in output can possibly differ from the one given in input as explained at the end of Sect. 7.2.3.

7.2.1 A Single Discrete Delay

As an extension to systems of Example 5.1 given in Sect. 5.1, consider the DDE

$$x'(t) = Ax(t) + Bx(t - \tau),$$

i.e.,

$$L\varphi = A\varphi(0) + B\varphi(-\tau), \quad \varphi \in X.$$

We choose the mesh of Chebyshev II nodes in $[-\tau, 0]$

$$\Omega_M = \left\{ \theta_{M,m}, \ m = 0, 1, \ldots, M \ : \ \theta_{M,m} = \frac{\tau}{2}\left(\cos\left(\frac{m\pi}{M}\right) - 1\right)\right\}. \quad (7.5)$$

It is immediate to verify that the first block-row of $\mathscr{A}_M$ in (7.4) has elements

$$a_j = \begin{cases} A & \text{if } j = 0, \\ 0_d & \text{if } j = 1, \ldots, M-1, \\ B & \text{if } j = M, \end{cases}$$

where 0_d denotes the $d \times d$ null matrix. This simply follows from

$$a_j = L(\ell_{M,j}(\cdot)I_d) = A\ell_{M,j}(0) + B\ell_{M,j}(-\tau) = A\ell_{M,j}(\theta_{M,0}) + B\ell_{M,j}(\theta_{M,M}),$$

for all $j = 0, 1, \ldots, M$ by applying the cardinal properties (5.3).

The remaining part of $\mathscr{A}_M$ in (7.4) is obtained from the so-called *Chebyshev differentiation matrix* [184], without the first row. The entries of this matrix are known explicitly and they can be built efficiently as shown in [184]. The subfunction difmat in eigAM.m (lines 218–234) is constructed indeed from [184].

The resulting matrix in $\mathbb{R}^{d(M+1) \times d(M+1)}$ reads

$$\mathscr{A}_M = \begin{pmatrix} A & 0_d & \cdots & 0_d & B \\ d_{1,0} & d_{1,1} & \cdots & d_{1,M-1} & d_{1,M} \\ \vdots & \vdots & & \vdots & \vdots \\ d_{M,0} & d_{M,1} & \cdots & d_{M,M-1} & d_{M,M} \end{pmatrix}$$

and it can be obtained with few MATLAB lines:

```
Id=eye(d);
OmegaM=tau*cos((0:M)*pi/M)-1)/2;
AM=kron(difmat(OmegaM),Id);
AM(1:d,:)=[A,zeros(d,d*(M-1)),B];
```

Note the use of the command kron for the Kronecker product (see Sect. 5.1).

7.2.2 A Single Distributed Delay

Let us consider the system

$$x'(t) = Ax(t) + \int_{-\tau}^{0} C(\theta)x(t+\theta)\,d\theta,$$

i.e.,

$$L\varphi = A\varphi(0) + \int_{-\tau}^{0} C(\theta)\varphi(\theta)\,d\theta, \quad \varphi \in X.$$

We choose Ω_M as in (7.5). The differentiation part of $\mathscr{A}_M$ in (7.4) is obtained as explained in the previous section. As for the elements of the first block-row, we get

$$a_j = L(\ell_{M,j}(\cdot)I_d) = A\ell_{M,j}(0) + \int_{-\tau}^{0} C(\theta)\ell_{M,j}(\theta)\,d\theta, \quad j = 0, 1, \ldots, M. \quad (7.6)$$

As anticipated in Sect. 5.3.3, the distributed delay term cannot be integrated analytically in general, hence we resort to a quadrature rule. Referring to the interval $[-1, 1]$ (as usual in the theory of numerical integration [184, 186]), we write

$$\int_{a}^{b} f(\theta)\,d\theta = \frac{b-a}{2} \int_{-1}^{1} f(\theta_{[a,b]}(z))\,dz \approx \frac{b-a}{2} \sum_{m=0}^{M} w_{M,m} f(\theta_{[a,b]}(z_{M,m})) \quad (7.7)$$

for a general integral, with the change of variable from $z \in [-1, 1]$ to $\theta \in [a, b]$

$$\theta_{[a,b]}(z) = \frac{b-a}{2}z + \frac{a+b}{2}$$

and where $z_{M,m}$ and $w_{M,m}$, $m = 0, 1, \ldots, M$, are, respectively, the $M+1$ nodes and weights of the chosen quadrature formula relevant to $[-1, 1]$.

Going back to the integral term in (7.6), one soon realizes that it is convenient to choose the quadrature formula based on the $M+1$ Chebyshev II nodes in $[-1, 1]$, known as Clenshaw–Curtis formula [184, 186]. In fact, being $z_{M,m} = \cos\left(\frac{m\pi}{M}\right)$, $m = 0, 1, \ldots, M$, it follows that $\theta_{[a,b]}(z_{M,m}) = \theta_{M,m}$ as in (7.5) and, for all $j = 0, 1, \ldots, M$,

$$\int_{-\tau}^{0} C(\theta)\ell_{M,j}(\theta)\,d\theta \approx \frac{\tau}{2} \sum_{m=0}^{M} w_{M,m} C(\theta_{M,m})\ell_{M,j}(\theta_{M,m}) = \frac{\tau}{2} w_{M,j} C(\theta_{M,j}),$$

7.2 The Infinitesimal Generator Approach

where we used (5.3) again. According to Sect. 5.3.3, we then set

$$L_M \varphi := A\varphi(0) + \sum_{m=0}^{M} w_{M,m} C(\theta_{M,m}) \varphi(\theta_{M,m}), \quad \varphi \in X,$$

and, consequently,

$$a_j = L_M(\ell_{M,j}(\cdot)I_d) = A\ell_{M,j}(0) + \frac{\tau}{2} w_{M,j} C(\theta_{M,j})$$
$$= \begin{cases} A + \dfrac{\tau}{2} w_{M,0} C(\theta_{M,0}) & \text{if } j = 0, \\ \dfrac{\tau}{2} w_{M,j} C(\theta_{M,j}) & \text{if } j = 1, \ldots, M. \end{cases}$$

The resulting matrix in $\mathbb{R}^{d(M+1) \times d(M+1)}$ reads

$$\mathscr{A}_M = \begin{pmatrix} A + \dfrac{\tau}{2} w_{M,0} C(\theta_{M,0}) & \dfrac{\tau}{2} w_{M,1} C(\theta_{M,1}) & \cdots & \dfrac{\tau}{2} w_{M,M} C(\theta_{M,M}) \\ d_{1,0} & d_{1,1} & \cdots & d_{1,M} \\ \vdots & \vdots & & \vdots \\ d_{M,0} & d_{M,1} & \cdots & d_{M,M} \end{pmatrix} \quad (7.8)$$

and it can be obtained with few MATLAB lines:

```
Id=eye(d);
OmegaM=tau*cos((0:M)*pi/M)-1)/2;
wqM=quadwei(M);
AM=kron(difmat(OmegaM),Id);
AM(1:d,:)=[A,zeros(d,d*M)]+tau/2*wqM.*C(OmegaM);
```

The subfunction `quadwei` in `eigAM.m` (lines 236–264) furnishes the vector `wqM` of the weights of the Clenshaw–Curtis quadrature. It is constructed from [184].

Let us recall that by proceeding this way, the quadrature error (5.33) in the final convergence result has the same decay behavior of the overall error in the approximation of the eigenvalues, i.e., it is spectrally accurate for a kernel function C of class C^∞, [184–186]. Moreover, no Lagrange coefficient needs to be evaluated, thus saving computational time.

7.2.3 The Piecewise Method

Let us consider the DDE generated by the general functional (7.3). If we choose the usual mesh (7.5), every discrete delay term in (7.3) whose delay does not coincide with a mesh node has to be interpolated. This is possible and the error is considered in Proposition 5.1 (see also Sect. 5.4). Instead, the treatment of each distributed delay term requires an appropriate quadrature whose nodes are not related in general to the nodes of Ω_M in (7.5). The values of the integrand at these quadrature nodes have to be reconstructed by interpolation, leading to an increase of computational cost. Notice that a single quadrature on $[-\tau, 0]$ would not be neither efficient nor convergent in the presence of discontinuities between the kernel functions C_k, $k = 1, \ldots, p$, at the inner delays τ_k, $k = 1, \ldots, p - 1$. Therefore, for given positive integers M_k, $k = 1, \ldots, p$, it is convenient to introduce the piecewise mesh

$$\Omega_M = \bigcup_{k=1}^{p} \Omega_{M_k}^{(k)}, \tag{7.9}$$

where $M := \sum_{k=1}^{p} M_k$ and

$$\Omega_{M_k}^{(k)} = \left\{ \theta_{M_k,m}^{(k)}, m = 0, 1, \ldots, M_k : \theta_{M_k,m}^{(k)} = \frac{\tau_k - \tau_{k-1}}{2} \cos\left(\frac{m\pi}{M_k}\right) - \frac{\tau_k + \tau_{k-1}}{2} \right\} \tag{7.10}$$

is the mesh of $M_k + 1$ Chebyshev II nodes discretizing $[-\tau_k, -\tau_{k-1}]$. Observe that

$$\theta_{M_k, M_k}^{(k)} = -\tau_k = \theta_{M_{k+1}, 0}^{(k+1)}, \quad k = 1, \ldots, p - 1, \tag{7.11}$$

i.e., the last node of a mesh is the first node of the next one. They correspond to the inner delays, Fig. 7.1. The mesh is constructed in the lines 28–36. However, note that this implementation is more general than the piecewise mesh used in Sect. 5.2: there, for simplicity of notation, every delay interval is discretized with the same number of nodes. See also the end of this section for further remarks.

With reference to the above mesh, we introduce the space $X_M = \mathbb{R}^{d(M+1)}$ as the discrete counterpart of X. A function $\varphi \in X$ is discretized by the vector $\Phi \in X_M$ of elements $\Phi = \left(\Phi_0^{(1)}, \Phi_1^{(1)}, \ldots, \Phi_{M_1}^{(1)}, \Phi_1^{(2)}, \ldots, \Phi_{M_2}^{(2)}, \ldots, \Phi_1^{(p)}, \ldots, \Phi_{M_p}^{(p)} \right)^T$ with $\Phi_m^{(k)} = \varphi(\theta_{M_k,m}^{(k)})$, $m = 0, 1, \ldots, M_k$, $k = 1, \ldots, p$, also recalling (7.11).

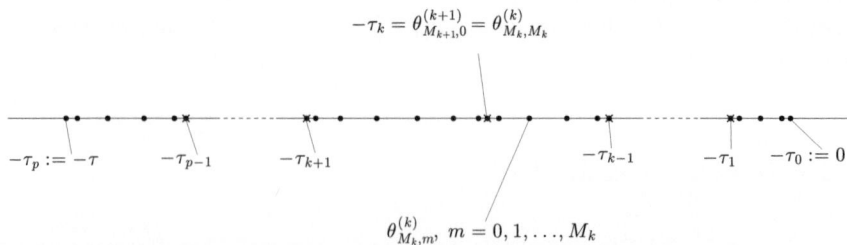

Fig. 7.1 An example of piecewise mesh (7.9): nodes (7.10) (•) and superposition nodes (7.11) (×)

7.2 The Infinitesimal Generator Approach

The value of the discretized infinitesimal generator on $\Phi \in X_M$ is given by

$$[\mathscr{A}_M \Phi]_m^{(k)} = \begin{cases} Ap_M(0) + \sum_{h=1}^{p} B_h p_M(-\tau_h) + \sum_{h=1}^{p} \int_{-\tau_h}^{-\tau_{h-1}} C_h(\theta) p_M(\theta) d\theta & \text{if } m = 0 \text{ and } k = 1, \\ p'_M(\theta_{M_k,m}^{(k)}) & \text{if } m = 1, \ldots, M_k \text{ and } k = 1, \ldots, p, \end{cases}$$

where p_M is the piecewise polynomial interpolating the elements of Φ on the nodes of the mesh Ω_M in (7.9). In particular, for all $k = 1, \ldots, p$, the restriction $p_{M_k}^{(k)}$ of p_M to $[-\tau_k, -\tau_{k-1}]$ is the unique M_k-degree polynomial that interpolates $\left(\Phi_{M_k,0}^{(k)}, \Phi_{M_k,1}^{(k)}, \ldots, \Phi_{M_k,M_k}^{(k)}\right)^T$ on the nodes of the mesh $\Omega_{M_k}^{(k)}$ in (7.10). By using the Lagrange representation relevant to each delay interval, we obtain the matrix

$$\mathscr{A}_M = \begin{pmatrix} a_0^{(1)} & a_1^{(1)} & \cdots & a_{M_1}^{(1)} & a_1^{(2)} & \cdots & a_{M_2}^{(2)} & \cdots & a_{M_{p-1}}^{(p-1)} & a_1^{(p)} & \cdots & a_{M_p}^{(p)} \\ d_{1,0}^{(1)} & d_{1,1}^{(1)} & \cdots & d_{1,M_1}^{(1)} & & & & & & & & \\ \vdots & \vdots & \ddots & \vdots & & & & & & & & \\ d_{M_1,0}^{(1)} & d_{M_1,1}^{(1)} & \cdots & d_{M_1,M_1}^{(1)} & & & & & & & & \\ & & & & d_{1,0}^{(2)} & d_{1,1}^{(2)} & \cdots & d_{1,M_2}^{(2)} & & & & \\ & & & & \vdots & \vdots & \ddots & \vdots & & & & \\ & & & & d_{M_2,0}^{(2)} & d_{M_2,1}^{(2)} & \cdots & d_{M_2,M_2}^{(2)} & & & & \\ & & & & & & & & \ddots & & & \\ & & & & & & & & & d_{1,0}^{(p)} & d_{1,1}^{(p)} & \cdots & d_{1,M_p}^{(p)} \\ & & & & & & & & & \vdots & \vdots & \ddots & \vdots \\ & & & & & & & & & d_{M_p,0}^{(p)} & d_{M_p,1}^{(p)} & \cdots & d_{M_p,M_p}^{(p)} \end{pmatrix}$$

in $\mathbb{R}^{d(M+1) \times d(M+1)}$, where missing entries are 0_d.

According to what explained in Sect. 7.2.1, for the differentiation part we have

$$d_{m,j}^{(k)} = \left(\ell_{M_k,j}^{(k)}\right)'\left(\theta_{M_k,m}^{(k)}\right) I_d, \ m = 1, \ldots, M_k, \ j = 0, 1, \ldots, M_k, \ k = 1, \ldots, p,$$

where $\ell_{M_k,0}^{(k)}, \ell_{M_k,1}^{(k)}, \ldots, \ell_{M_k,M_k}^{(k)}$ are the Lagrange coefficients relevant to the mesh $\Omega_{M_k}^{(k)}$ in (7.10) discretizing $[-\tau_k, -\tau_{k-1}]$. In eigAM.m, this is implemented in a for loop along the delays (lines 38–47) where, again, the subfunction difmat is used for the Chebyshev differentiation matrix relevant to the nodes in each delay interval.

As for the first block-row, instead, it is not difficult to see from Sect. 7.2.2 that

$$a_j^{(k)} = \begin{cases} A + \dfrac{\tau_1}{2} w_{M_1,0}^{(1)} C_1(\theta_{M_1,0}^{(1)}) & \text{if } j = 0 \text{ and } k = 1, \\ \dfrac{\tau_k - \tau_{k-1}}{2} w_{M_k,j}^{(k)} C_k(\theta_{M_k,j}^{(k)}) & \text{if } j = 1, \ldots, M_k - 1 \text{ and } k = 1, \ldots, p, \\ B_k + \dfrac{\tau_k - \tau_{k-1}}{2} w_{M_k,M_k}^{(k)} C_k(\theta_{M_k,M_k}^{(k)}) \\ \quad + \dfrac{\tau_{k+1} - \tau_k}{2} w_{M_{k+1},0}^{(k+1)} C_{k+1}(\theta_{M_{k+1},0}^{(k+1)}) & \text{if } j = M_k \text{ and } k = 1, \ldots, p-1, \\ B_p + \dfrac{\tau_p - \tau_{p-1}}{2} w_{M_p,M_p}^{(p)} C_p(\theta_{M_p,M_p}^{(p)}) & \text{if } j = M_k \text{ and } k = p, \end{cases}$$

where $w_{M_k,m}^{(k)}$, $m = 0, 1, \ldots, M_k$, are the quadrature weights of the Clenshaw–Curtis formula for $M_k + 1$ nodes in $[-1, 1]$, $k = 1, \ldots, p$. In the code eigAM.m, this part is implemented in the lines 49–66 in a for loop along the delays where, again, the subfunction quadwei is used to obtain the quadrature weights. Notice that this is necessary since, in general, M_k can vary with k.

About the last point, one can choose the number of nodes in each delay interval in several ways. Assume that the objective is to ensure a desired final accuracy at the minimum computational cost. Based on the convergence result stated in Theorem 5.2 and on the comments in Sect. 5.4, it is not difficult to argue that (after rounding) a good choice is $M_1, \ldots, M_k$ satisfying

$$\begin{cases} \left(\dfrac{\tau_1 - \tau_0}{M_1}\right)^{M_1} = \cdots = \left(\dfrac{\tau_p - \tau_{p-1}}{M_p}\right)^{M_p} \\ \sum_{k=1}^{p} M_k = M \end{cases}$$

for a given positive integer M and given delays. This system of nonlinear equations is solved (numerically) by the subfunction em1 in eigAM.m (lines 165–192). The alternative em2 (lines 194–216) solves the similar problem for given M points in the largest delay interval. The true number of nodes finally used can differ slightly from the input due to integer rounding. This is why it is given in output, mainly for the purpose of analyzing the convergence of the error w.r.t. increasing M.

Remark 7.2 The question about what is the minimum discretization index M ensuring a prescribed tolerance on the approximated roots is legitimate. In the case of the IG approach with pseudospectral differentiation methods, a first (partially heuristic) answer is given in [205] for DDEs (7.1) without distributed delay terms. There, the authors estimate such value of M by analyzing the properties of the interpolation of the exponential function on the complex plane. This strategy enables to guarantee a prescribed error on the roots in a half plane bounded to the left (always finitely

7.2 The Infinitesimal Generator Approach

many according to Proposition 3.4). An alternative way consists in estimating all the constants in the error bound (5.30). A first attempt in this direction is done in [149], by using the results of [108] to estimate the constant C_3 in Proposition 5.3. Note that the procedure is quite involved w.r.t. numerical methods with convergence of finite order as Runge–Kutta or Linear Multistep methods [30, 31]. In the latter case, a simple Richardon's extrapolation can be applied. Nevertheless, it should be stressed that, given the rapid convergence of the pseudospectral differentiation method, the discretization index M is usually low and an order of tenths is in general sufficient to reach machine precision (as, indeed, in Chap. 8).

7.3 The Solution Operator Approach

As explained in Sect. 7.1, the implementation of the SO approach with pseudospectral collocation methods developed in Chap. 6 is referred to the prototype model (7.1) for the nonautonomous case, i.e., with coefficients depending on the time t, in particular periodically. The latter is obtained from (2.6) by choosing

$$L(t)\varphi = A(t)\varphi(0) + \sum_{k=1}^{p} B_k(t)\varphi(-\tau_k) + \sum_{k=1}^{p} \int_{-\tau_k}^{-\tau_{k-1}} C_k(t,\theta)\varphi(\theta)\,d\theta, \quad \varphi \in X. \tag{7.12}$$

Let us remark that (7.1) can describe a DDE which is either originally linear and periodic in the considered application or it is the result of the linearization of a nonlinear model around a periodic solution. Then the approximation of the characteristic multipliers gives information on the global stability of the zero solution in the former case or on the local stability of the concerned periodic solution in the latter case (through the Principle of Linearized Stability, Theorem 4.3). The pseudospectral collocation method presented in Chap. 6 is suitable for discretizing a general evolution operator $T(s+h, s)$ for $s \in \mathbb{R}$ and $h > 0$. Therefore, we consider (7.12) with general nonautonomous coefficients, not necessarily periodic.

At the end of Sect. 6.1.3, it is left suspended how to recover the matrix representation of the discretized operator $T_{M,N}$ defined through (6.9) and (6.10). To this aim, observe that by (6.13), for any $\Phi \in X_M$ and for $Z^* \in X_N^+$ the unique solution of (6.10), we can rewrite (6.9) as

$$T_{M,N}\Phi = T_M^{(1)}\Phi + T_{M,N}^{(2)} Z^* \tag{7.13}$$

where $T_M^{(1)} : X_M \to X_M$ is given by

$$T_M^{(1)}\Phi = R_M(V_1 P_M \Phi)_h, \quad \Phi \in X_M, \tag{7.14}$$

and $T_{M,N}^{(2)} : X_N^+ \to X_M$ is given by

$$T_{M,N}^{(2)} Z = R_M(V_2 P_N^+ Z)_h, \quad Z \in X_N^+. \tag{7.15}$$

By using (6.13) again, (6.10) can be rewritten as

$$\left(I_{X_N^+} - U_N^{(2)}\right) Z^* = U_{M,N}^{(1)} \Phi, \tag{7.16}$$

where $U_{M,N}^{(1)} : X_M \to X_N^+$ is given by

$$U_{M,N}^{(1)} \Phi = R_N^+ \mathscr{F}_s V_1 P_M \Phi, \quad \Phi \in X_M, \tag{7.17}$$

and $U_N^{(2)} : X_N^+ \to X_N^+$ is given, for $\mathscr{F}_s$ as defined in (6.2), by

$$U_N^{(2)} Z = R_N^+ \mathscr{F}_s V_2 P_N^+ Z, \quad Z \in X_N^+. \tag{7.18}$$

For the definitions of V_1, V_2, X_M, X_N^+, R_M, P_M, R_N^+ and P_N^+ go back to the beginning of Chap. 6. Let us note that $T_M^{(1)}$ and $T_{M,N}^{(2)}$ are independent of the model coefficients, depending only on the step h and on the delays τ_k, $k = 1, \ldots, p$. Instead, $U_{M,N}^{(1)}$ and $U_N^{(2)}$ depend on $\mathscr{F}_s$ which, from (6.2) and (7.12) and for $x \in X^{\pm}$ and $t \in [0, h]$, reads

$$(\mathscr{F}_s x)(t) = A(s+t)x(t) + \sum_{k=1}^{p} B_k(s+t)x(t-\tau_k) + \sum_{k=1}^{p} \int_{-\tau_k}^{-\tau_{k-1}} C_k(s+t, \theta)x(t+\theta)\, d\theta. \tag{7.19}$$

Now, under the conditions C1 and C2 of Theorem 6.1, for $N \geq N_0$ (N_0 given in there), the operator $I_{X_N^+} - U_N^{(2)}$ in (7.16) is invertible (recall Proposition 6.1). Therefore, the finite dimensional operator $T_{M,N}$ in (7.13) can be expressed by

$$T_{M,N} = T_M^{(1)} + T_{M,N}^{(2)} \left(I_{X_N^+} - U_N^{(2)}\right)^{-1} U_{M,N}^{(1)}. \tag{7.20}$$

Finally, by identifying the space X_N^+ with $\mathbb{R}^{dN}$ and the space X_M with $\mathbb{R}^{d(M+1)}$ for $h \geq \tau$ or with $\mathbb{R}^{d(QM+1)}$ for $h < \tau$, we can consider (7.20) as the matrix representation in the canonical basis of the corresponding operator. Its eigenvalues (computed by the standard methods for the computation of matrix eigenvalues) are the approximation of the eigenvalues of $T(s+h, s)$ as proved in Chap. 6. Moreover,

7.3 The Solution Operator Approach

by (7.20), the matrix $T_{M,N}$ can be recovered by constructing the matrices representing the above finite dimensional operators $T_M^{(1)}$, $T_{M,N}^{(2)}$, $U_{M,N}^{(1)}$, $U_N^{(2)}$, respectively. In the following sections, we explain in detail how to recover these matrices and how their construction is implemented in the MATLAB function `eigTMN.m`.

The overall structure of `eigTMN.m` is organized as follows:

- the main comment (lines 1–24);
- the conversion of (7.2) as defined in `myDDE.m` into (7.1) (lines 26–29);
- the definition of the meshes Ω_M and Ω_N^+, see Sect. 7.3.1 (lines 31–50);
- the construction of the matrix $T_M^{(1)}$ (lines 52–71);
- the construction of the matrix $T_{M,N}^{(2)}$ (lines 73–107);
- the construction of the matrix $U_{M,N}^{(1)}$ (lines 109–809);
- the construction of the matrix $U_N^{(2)}$ (lines 811–900);
- the construction of the matrix $T_{M,N}$ (lines 902–903);
- the computation of the eigenvalues μ of $T_{M,N}$ by `eig`, ordered by decreasing modulus (lines 905–907);
- a set of auxiliary functions, see later on (lines 909–1,048).

The standard call is

`mu=eigTMN('myDDE',par,s,h,M,N)`

where the inputs are the name of the script `myDDE.m`, a vector `par` of possible parameter values as explained in Sect. 7.1, the starting time s, the step h, and the positive integers M and N defining the discretization indices. The output is the vector of eigenvalues of $T_{M,N}$, which are $d(M+1)$ if $h \geq \tau$ and $d(QM+1)$ if $h < \tau$.

7.3.1 The Meshes

In the MATLAB implementation of the SO approach with the pseudospectral collocation method, we use meshes of Chebyshev-type nodes. According to C2 in Theorem 6.1, they ensure the convergence of infinite order as stated in Theorem 6.6. For later convenience, we express the various meshes as functions of the nodes in $[-1, 1]$. Therefore, we introduce, for a positive integer N, the N Chebyshev I nodes in $[-1, 1]$

$$z_{N,n}^{(I)} = \cos\left(\frac{(2n-1)\pi}{2N}\right), \quad n = 1, \ldots, N, \tag{7.21}$$

and, for a positive integer M, the $M+1$ Chebyshev II nodes in $[-1, 1]$

$$z_{M,m}^{(II)} = \cos\left(\frac{m\pi}{M}\right), \quad m = 0, 1, \ldots, M. \tag{7.22}$$

note that both sets of nodes are ordered from right to left. Moreover, we recall from Sect. 7.2.2 that for any $a, b \in \mathbb{R}$, $a < b$,

$$\theta_{[a,b]}(z) = \frac{b-a}{2} z + \frac{a+b}{2} \tag{7.23}$$

is the change of variable from $z \in [-1, 1]$ to $\theta \in [a, b]$ and

$$z_{[a,b]}(\theta) = \frac{2\theta - b - a}{b - a} \tag{7.24}$$

is the change of variable from $\theta \in [a, b]$ to $z \in [-1, 1]$.

If $h \geq \tau$, the interval $[-\tau, 0]$ is discretized with $M + 1$ Chebyshev II nodes. Therefore, we use the mesh Ω_M in (7.5), i.e.,

$$\Omega_M = \left\{ \theta_{M,m}, \ m = 0, 1, \ldots, M \ : \ \theta_{M,m} = \frac{\tau}{2} \left(z_{M,m}^{(II)} - 1 \right) \right\}, \tag{7.25}$$

where the nodes are ordered right to left.

Instead, if $h < \tau$, by recalling from Sect. 6.1.1.2, we set $Q := \min\{q \in \mathbb{N} : qh \geq \tau\}$ (note that $Q > 1$) and discretize $[-\tau, 0]$ with the piecewise mesh

$$\Omega_M := \bigcup_{q=1}^{Q} \Omega_M^{(q)}, \tag{7.26}$$

where, for $q = 1, \ldots, Q - 1$,

$$\Omega_M^{(q)} = \left\{ \theta_{M,m}^{(q)}, \ m = 0, 1, \ldots, M \ : \ \theta_{M,m}^{(q)} = h \left(\frac{z_{M,m}^{(II)} + 1}{2} - q \right) \right\} \tag{7.27}$$

is the mesh of $M + 1$ Chebyshev II nodes in $[-qh, -(q-1)h]$ while, for $q = Q$,

$$\Omega_M^{(Q)} = \left\{ \theta_{M,m}^{(Q)}, \ m = 0, 1, \ldots, M \ : \ \theta_{M,m}^{(Q)} = \frac{\tau - (Q-1)h}{2} z_{M,m}^{(II)} - \frac{\tau + (Q-1)h}{2} \right\} \tag{7.28}$$

is the mesh of $M + 1$ Chebyshev II nodes in the last interval $[-\tau, -(Q-1)h]$. Note the superposition

$$\theta_{M,M}^{(q)} = -qh = \theta_{M,0}^{(q+1)}, \quad q = 1, \ldots, Q - 1, \tag{7.29}$$

Figure 7.2. Observe also that we use the same number of nodes in each interval since there is no reason a priori to do otherwise (as it were the case in Sect. 7.2.3). Moreover, all the nodes and all the intervals are ordered right to left.

7.3 The Solution Operator Approach

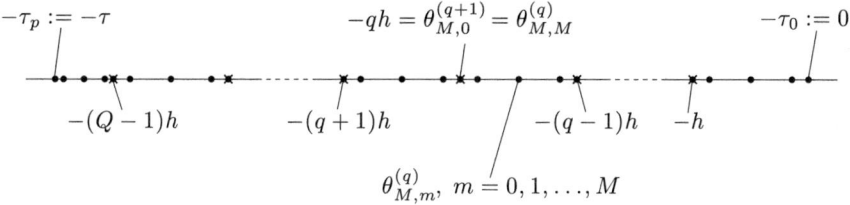

Fig. 7.2 An example of piecewise mesh (7.26): nodes (7.27) and (7.28) (•) and superposition nodes (7.29) (×)

The interval $[0, h]$ is discretized with N Chebyshev I nodes. Therefore, we use

$$\Omega_N^+ = \left\{ t_{N,n}, \ n = 1, \ldots, N \ : \ t_{N,n} = \frac{h}{2}\left(1 - z_{N,n}^{(I)}\right) \right\}, \tag{7.30}$$

where the nodes are now ordered left to right.

In eigTMN.m, the meshes Ω_M discretizing $[-\tau, 0]$ are implemented in the lines 31–45. In particular, for $h < \tau$ the mesh (7.26) is stored in the matrix

$$\Omega_M = \begin{pmatrix} \theta_{M,0}^{(1)} & \theta_{M,1}^{(1)} & \cdots & \theta_{M,M}^{(1)} \\ \theta_{M,0}^{(2)} & \theta_{M,1}^{(2)} & \cdots & \theta_{M,M}^{(2)} \\ \vdots & \vdots & \ddots & \vdots \\ \theta_{M,0}^{(Q)} & \theta_{M,1}^{(Q)} & \cdots & \theta_{M,M}^{(Q)} \end{pmatrix} \in \mathbb{R}^{Q \times (M+1)},$$

where the qth row contains the nodes of the mesh $\Omega_M^{(q)}$ in (7.27) for $q = 1, \ldots, Q-1$ and the nodes of the mesh $\Omega_M^{(Q)}$ in (7.28) for $q = Q$. The mesh Ω_N^+ discretizing $[0, h]$ is implemented in the lines 47–50. The integers M and N are given in input.

7.3.2 The Matrix $T_M^{(1)}$

In (7.14), we have $(V_1 P_M \Phi)_h (\theta) = (V_1 P_M \Phi)(h + \theta)$, $\theta \in [-\tau, 0]$, where, according to (6.14) and (6.1),

$$(V_1 P_M \Phi)(t) = \begin{cases} (P_M \Phi)(0) & \text{if } t \in [0, h], \\ (P_M \Phi)(t) & \text{if } t \in [-\tau, 0]. \end{cases} \tag{7.31}$$

If $h \geq \tau$, P_M and R_M in (7.14) are based on the mesh Ω_M in (7.25). Therefore, $\Phi = (\Phi_0, \Phi_1, \ldots, \Phi_M)^T \in \mathbb{R}^{d(M+1)}$ and it is immediate to see that for all $\theta \in [-\tau, 0]$, $(V_1 P_M \Phi)_h (\theta) = \Phi_0$. Consequently, the resulting matrix in $\mathbb{R}^{d(M+1) \times d(M+1)}$ reads

$$T_M^{(1)} = \begin{pmatrix} 1 & 0 & \cdots & 0 \\ 1 & 0 & \cdots & 0 \\ \vdots & \vdots & \ddots & \vdots \\ 1 & 0 & \cdots & 0 \end{pmatrix} \otimes I_d.$$

This matrix is easily constructed in the single line 54 of `eigTMN.m`.

If $h < \tau$, P_M, and R_M in (7.14) are based on the piecewise mesh Ω_M (7.26). Hence,

$$\Phi = \left(\Phi_0^{(1)}, \ldots, \Phi_{M-1}^{(1)}, \Phi_0^{(2)}, \ldots, \Phi_{M-1}^{(2)}, \ldots, \Phi_0^{(Q)}, \ldots, \Phi_{M-1}^{(Q)}, \Phi_M^{(Q)} \right)^T \in \mathbb{R}^{d(QM+1)}, \tag{7.32}$$

according also to the superposition (7.29). Since

$$(V_1 P_M \Phi)_h (\theta) = \begin{cases} (P_M \Phi)(0) & \text{if } \theta \in [-h, 0], \\ (P_M \Phi)(h + \theta) & \text{if } \theta \in [-\tau, -h], \end{cases}$$

when we apply R_M in front, we get $[R_M (V_1 P_M \Phi)_h]_m^{(1)} = \Phi_0^{(1)}$, $m = 0, 1, \ldots, M-1$, while $[R_M (V_1 P_M \Phi)_h]_m^{(q)} = (P_M \Phi)\left(h + \theta_{M,m}^{(q)}\right)$, $m = 0, 1, \ldots, M-1$, for all $q = 2, \ldots, Q$. Given (7.27), it follows that $h + \theta_{M,m}^{(q)} = \theta_{M,m}^{(q-1)}$ for all $q = 2, \ldots, Q-1$, which leads to $[R_M (V_1 P_M \Phi)_h]_m^{(q)} = \Phi_m^{(q-1)}$, $m = 0, 1, \ldots, M-1$. When $q = Q$, instead, every point $h + \theta_{M,m}^{(Q)}$ falls in the interval $[-(Q-1)h, -(Q-2)h]$, but since in general the Qth interval is shorter than the $Q-1$st one, interpolation must be performed. Hence, $[R_M (V_1 P_M \Phi)_h]_m^{(Q)} = \sum_{j=0}^{M} \ell_{M,j}^{(Q-1)} \left(h + \theta_{M,m}^{(Q)} \right) \Phi_j^{(Q-1)}$, $m = 0, 1, \ldots, M$, with $\ell_{M,0}^{(Q-1)}, \ell_{M,1}^{(Q-1)}, \ldots, \ell_{M,M}^{(Q-1)}$, the Lagrange coefficients relevant to the mesh $\Omega_M^{(Q-1)}$ in (7.27). Therefore, by setting

$$t_{m,j}^{(Q)} = \ell_{M,j}^{(Q-1)} \left(h + \theta_{M,m}^{(Q)} \right), \quad m, j = 0, 1, \ldots, M, \tag{7.33}$$

the resulting matrix in $\mathbb{R}^{d(QM+1) \times d(QM+1)}$ reads

7.3 The Solution Operator Approach

$$T_M^{(1)} = \begin{pmatrix} 1 & & & & & & & & & & \\ & \ddots & & & & & & & & & \\ & & 1 & & & & & & & & \\ & & & 1 \cdots 0 & & & & & & & \\ & & & \vdots \ddots \vdots & & & & & & & \\ & & & 0 \cdots 1 & & & & & & & \\ & & & & \ddots & & & & & & \\ & & & & & 1 \cdots 0 & & & & & \\ & & & & & \vdots \ddots \vdots & & & & & \\ & & & & & 0 \cdots 1 & & & & & \\ & & & & & & t_{0,0}^{(Q)} & \cdots & t_{0,M-1}^{(Q)} & t_{0,M}^{(Q)} & 0 \cdots 0 \\ & & & & & & \vdots & \ddots & \vdots & \vdots & \vdots \ddots \vdots \\ & & & & & & t_{M-1,0}^{(Q)} & \cdots & t_{M-1,M-1}^{(Q)} & t_{M-1,M}^{(Q)} & 0 \cdots 0 \\ & & & & & & t_{M,0}^{(Q)} & \cdots & t_{M,M-1}^{(Q)} & t_{M,M}^{(Q)} & 0 \cdots 0 \end{pmatrix} \otimes I_d,$$

(7.34)

where missing entries are zeroes. To avoid confusion, the matrix above is represented in blocks according to the enumeration $q = 1, \ldots, Q$ with reference to the rows and to the ordering (7.32). For $q = 1, \ldots, Q - 1$ each block has size $M \times M$ and its entries are indexed from 0 to $M - 1$; for $q = Q$ the block has size $(M+1) \times (M+1)$ and its entries are indexed from 0 to M. Note that if Q satisfies $Qh = \tau$, the above matrix reduces to a simple shift plus constant extension (of $\Phi_0^{(1)}$), thanks to (5.3) as applied to (7.33). If $Qh > \tau$, instead, (7.33) requires the evaluation of the Lagrange coefficients at different points, performed as explained in the forthcoming section (meshes more general than (7.27) and (7.28) do not alter the structure of (7.34), but for the blocks in the lower diagonal that are no more identities [44]).

7.3.2.1 Barycentric Lagrange Interpolation

The matrix (7.34) is built in the lines 56–69. Let us focus on the evaluation of the Lagrange coefficients in (7.33). For efficiency we use the barycentric interpolation, [24]. Since used intensively in the sequel, we summarize it here for general Lagrange coefficients $\ell_0, \ell_1, \ldots, \ell_M$ relevant to a mesh of nodes $\theta_0, \theta_1, \ldots, \theta_M$ discretizing an interval $[a, b]$. First, it is convenient to transform $\theta \in [a, b]$ into $z \in [-1, 1]$ and vice versa through (7.23) and (7.24). Accordingly, let $z_0, z_1, \ldots, z_M$ be the corresponding nodes in $[-1, 1]$, i.e., $\theta_m = \theta_{[a,b]}(z_m)$, $m = 0, 1, \ldots, M$, and observe that

$$\ell_j(\theta_{[a,b]}(z)) = \prod_{\substack{m=0\\m\neq j}}^{M} \frac{\theta_{[a,b]}(z) - \theta_m}{\theta_j - \theta_m} = \prod_{\substack{m=0\\m\neq j}}^{M} \frac{z - z_m}{z_j - z_m} = \bar{\ell}_j(z),$$

$$j = 0, 1, \ldots, M, \quad z \in [-1, 1],$$

with $\bar{\ell}_0, \bar{\ell}_1, \ldots, \bar{\ell}_M$ the Lagrange coefficients relevant to the nodes in $[-1, 1]$.

Let the *barycentric weights* of interpolation associated to the nodes in $[-1, 1]$ be

$$w_j := \frac{1}{\prod_{\substack{m=0\\m\neq j}}^{M} (z_j - z_m)}, \quad j = 0, 1, \ldots, M.$$

They depend only on the nodes and they can be computed efficiently in $O(M^2)$ flops, see the algorithm in [24], which is the core of the subfunction `barywei` (lines 1,001–1,018). Define also the *nodal polynomial* relevant to the same nodes as

$$\pi_{M+1}(z) := \prod_{m=0}^{M} (z - z_m), \quad z \in [-1, 1].$$

It can be computed in just $O(M)$ flops. Therefore, since

$$\bar{\ell}_j(z) = \frac{w_j}{z - z_j} \cdot \pi_{M+1}(z), \quad j = 0, 1, \ldots, M, \quad z \in [-1, 1],$$

the change of the point of evaluation z has only linear cost since the barycentric weights are computed once for all. This is exactly how the Lagrange coefficients relevant to any interval $[a, b]$ are computed in `eigTMN.m` whenever required:

$$\ell_j(\theta) = \frac{w_j}{z_{[a,b]}(\theta) - z_j} \cdot \pi_{M+1}(z_{[a,b]}(\theta)), \quad j = 0, 1, \ldots, M, \quad \theta \in [a, b], \quad (7.35)$$

compare the lines 62–67 w.r.t. (7.33). The barycentric weights are computed in line 33 for Chebyshev II nodes (7.22) and in line 49 for Chebyshev I nodes (7.21).

7.3.3 The Matrix $T_{M,N}^{(2)}$

In (7.15), we have $\left(V_2 P_N^+ Z\right)_h (\theta) = \left(V_2 P_N^+ Z\right)(h + \theta)$, $\theta \in [-\tau, 0]$, where

$$\left(V_2 P_N^+ Z\right)(t) = \begin{cases} \int_0^t (P_N^+ Z)(s)\,ds & \text{if } t \in [0, h], \\ 0 & \text{if } t \in [-\tau, 0], \end{cases} \quad (7.36)$$

7.3 The Solution Operator Approach

according to (6.15) and (6.1) and

$$(P_N^+ Z)(t) = \sum_{n=1}^{N} \ell_{N,n}^+(t) Z_n, \quad t \in [0, h], \tag{7.37}$$

with $\ell_{N,1}^+, \ldots, \ell_{N,N}^+$ the Lagrange coefficients relevant to the mesh Ω_N^+ in (7.30).

If $h \geq \tau$, R_M in (7.15) is based on the mesh Ω_M (7.25). Therefore, being $h + \theta_{M,m} \geq 0$, for all $m = 0, 1, \ldots, M$, it is not difficult to obtain the matrix in $\mathbb{R}^{d(M+1) \times dN}$

$$T_{M,N}^{(2)} = \begin{pmatrix} \int_0^{h+\theta_{M,0}} \ell_{N,1}^+(t)\, dt & \cdots & \int_0^{h+\theta_{M,0}} \ell_{N,N}^+(t)\, dt \\ \vdots & & \vdots \\ \int_0^{h+\theta_{M,M}} \ell_{N,1}^+(t)\, dt & \cdots & \int_0^{h+\theta_{M,M}} \ell_{N,N}^+(t)\, dt \end{pmatrix} \otimes I_d.$$

This matrix is built in the lines 75–89. Each integral is computed via the Clenshaw–Curtis formula, i.e., (7.7) for the Chebyshev II nodes $z_{M,j}^{(II)}$ in (7.22) and the relevant quadrature weights $w_{M,j}^{(\text{quad})}$, $j = 0, 1, \ldots, M$, obtained with the subfunction quadwei (lines 1,020–1,048). The Lagrange coefficients at the corresponding nodes

$$\theta_{[0,h+\theta_{M,m}]}(z_{M,j}^{(II)}) = \frac{h + \theta_{M,m}}{2}(z_{M,j}^{(II)} + 1)$$

in $[0, h + \theta_{M,m}]$ for $m = 0, 1, \ldots, M$ are computed through the barycentric formula (7.35) relevant to the Chebyshev I nodes $z_{N,n}^{(I)}$ in (7.21), giving the evaluation points

$$z_{[0,h]}(\theta_{[0,h+\theta_{M,m}]}(z_{M,j}^{(II)})) = -\frac{2\theta_{[0,h+\theta_{M,m}]}(z_{M,j}^{(II)}) - h}{h} = 1 - \frac{(h + \theta_{M,m})(z_{M,j}^{(II)} + 1)}{h}.$$

The minus sign in the first equivalence is due to the reverse ordering between the Chebyshev I nodes (7.21) (right to left) and the mesh Ω_N^+ in (7.30) (left to right). The general (m, n)th entry of $T_{M,N}^{(2)}$ for $m = 0, 1, \ldots, M$ and $n = 1, \ldots, N$ is given by

$$\int_0^{h+\theta_{M,m}} \ell_{N,n}^+(t)\, dt \approx \frac{h + \theta_{M,m}}{2} \sum_{j=0}^{M} w_{M,j}^{(\text{quad})} \ell_{N,n}^+(\theta_{[0,h+\theta_{M,m}]}(z_{M,j}^{(II)}))$$

$$= w_{N,n}^{(\text{bary})} \cdot \frac{h + \theta_{M,m}}{2} \sum_{j=0}^{M} w_{M,j}^{(\text{quad})} \frac{\pi_{M+1}(z_{[0,h]}(\theta_{[0,h+\theta_{M,m}]}(z_{M,j}^{(II)})))}{z_{[0,h]}(\theta_{[0,h+\theta_{M,m}]}(z_{M,j}^{(II)})) - z_{N,n}^{(I)}}.$$

This is how integrals of Lagrange coefficients (and of more general integrands in the sequel) are computed in `eigTMN.m` and thus we show it explicitly at least once.

If $h < \tau$, R_M in (7.15) is based on the mesh Ω_M (7.26). Therefore, being

$$h + \theta_{M,m}^{(q)} \begin{cases} \geq 0 & \text{if } m = 0, 1, \ldots, M-1 \text{ and } q = 1, \\ < 0 & \text{if } m = 0, 1, \ldots, M \text{ and } q = 2, \ldots, Q, \end{cases}$$

we obtain the matrix in $\mathbb{R}^{d(QM+1)\times dN}$ (lines 91–105, similarly to the case $h \geq \tau$):

$$T_{M,N}^{(2)} = \begin{pmatrix} \int_0^{h+\theta_{M,0}^{(1)}} \ell_{N,1}^+(t)\,dt & \cdots & \int_0^{h+\theta_{M,0}^{(1)}} \ell_{N,N}^+(t)\,dt \\ \vdots & & \vdots \\ \int_0^{h+\theta_{M,M-1}^{(1)}} \ell_{N,1}^+(t)\,dt & \cdots & \int_0^{h+\theta_{M,M-1}^{(1)}} \ell_{N,N}^+(t)\,dt \\ 0 & \cdots & 0 \\ \vdots & & \vdots \\ 0 & \cdots & 0 \end{pmatrix} \otimes I_d.$$

7.3.4 The Matrix $U_{M,N}^{(1)}$

From (7.17) and (7.19) we have, according to the mesh Ω_N^+ (7.30),

$$[U_{M,N}^{(1)}\Phi]_n = A(s + t_{N,n})(V_1 P_M \Phi)(t_{N,n}) + \sum_{k=1}^{p} B_k(s + t_{N,n})(V_1 P_M \Phi)(t_{N,n} - \tau_k)$$
$$+ \sum_{k=1}^{p} \int_{-\tau_k}^{-\tau_{k-1}} C_k(s + t_{N,n}, \theta)(V_1 P_M \Phi)(t_{N,n} + \theta)\,d\theta, \quad n = 1, \ldots, N.$$

If $h \geq \tau$, P_M in (7.31) is based on the mesh Ω_M (7.25). We have to establish whether the arguments $t_{N,n} - \tau_k$ and $t_{N,n} + \theta$ fall in $[0, h]$, where $V_1 P_M \Phi$ has the constant value Φ_0, or in $[-\tau, 0]$, where $V_1 P_M \Phi$ coincides with $P_M \Phi$. Hence, let

$$\hat{N} := \begin{cases} 0 & \text{if } t_{N,n} \geq \tau \text{ for all } n = 1, \ldots, N, \\ \max\{n : t_{N,n} < \tau\} & \text{otherwise}, \end{cases}$$

7.3 The Solution Operator Approach

and

$$k(t) := \begin{cases} k & \text{if } \tau_k \leq t < \tau_{k+1}, \quad k = 0, 1, \ldots p-1, \\ p & \text{if } t = \tau, \end{cases} \quad (7.38)$$

for $t \in [0, \tau]$. It follows that

$$[U^{(1)}_{M,N}\Phi]_n = A(s + t_{N,n})\Phi_0 + \sum_{k=1}^{k(t_{N,n})} B_k(s + t_{N,n})\Phi_0$$

$$+ \sum_{k=k(t_{N,n})+1}^{p} B_k(s + t_{N,n}) \sum_{m=0}^{M} \ell_{M,m}(t_{N,n} - \tau_k)\Phi_m$$

$$+ \sum_{k=1}^{k(t_{N,n})} \int_{-\tau_k}^{-\tau_{k-1}} C_k(s + t_{N,n}, \theta)\Phi_0 \, d\theta + \int_{-t_{N,n}}^{-\tau_{k(t_{N,n})}} C_{k(t_{N,n})+1}(s + t_{N,n}, \theta)\Phi_0 \, d\theta$$

$$+ \int_{-\tau_{k(t_{N,n})+1}}^{-t_{N,n}} C_{k(t_{N,n})+1}(s + t_{N,n}, \theta) \sum_{m=0}^{M} \ell_{M,m}(t_{N,n} + \theta)\Phi_m \, d\theta$$

$$+ \sum_{k=k(t_{N,n})+2}^{p} \int_{-\tau_k}^{-\tau_{k-1}} C_k(s + t_{N,n}, \theta) \sum_{m=0}^{M} \ell_{M,m}(t_{N,n} + \theta)\Phi_m \, d\theta,$$

for $n = 1, \ldots, \hat{N}$ and

$$[U^{(1)}_{M,N}\Phi]_n = A(s+t_{N,n})\Phi_0 + \sum_{k=1}^{p} B_k(s+t_{N,n})\Phi_0 + \sum_{k=1}^{p} \int_{-\tau_k}^{-\tau_{k-1}} C_k(s+t_{N,n}, \theta)\Phi_0 \, d\theta,$$

for $n = \hat{N}+1, \ldots, N$. Consequently, the resulting matrix in $\mathbb{R}^{dN \times d(M+1)}$ reads

$$U^{(1)}_{M,N} = \begin{pmatrix} D_{1,0} + E_{1,0} & E_{1,1} & \cdots & E_{1,M} \\ \vdots & \vdots & & \vdots \\ D_{\hat{N},0} + E_{\hat{N},0} & E_{\hat{N},1} & \cdots & E_{\hat{N},M} \\ D_{\hat{N}+1,0} & 0 & \cdots & 0 \\ \vdots & \vdots & & \vdots \\ D_{N,0} & 0 & \cdots & 0 \end{pmatrix},$$

where

$$D_{n,0} = A(s + t_{N,n}) + \sum_{k=1}^{k(t_{N,n})} B_k(s + t_{N,n}) + \sum_{k=1}^{k(t_{N,n})} \int_{-\tau_k}^{-\tau_{k-1}} C_k(s + t_{N,n}, \theta)\,d\theta$$

$$+ \int_{-t_{N,n}}^{-\tau_{k(t_{N,n})}} C_{k(t_{N,n})+1}(s + t_{N,n}, \theta)\,d\theta, \quad n = 1, \ldots, \hat{N},$$

$$D_{n,0} = A(s + t_{N,n}) + \sum_{k=1}^{p} B_k(s + t_{N,n}) + \sum_{k=1}^{p} \int_{-\tau_k}^{-\tau_{k-1}} C_k(s + t_{N,n}, \theta)\,d\theta, \quad n = \hat{N} + 1, \ldots, N,$$

and

$$E_{n,m} = \sum_{k=k(t_{N,n})+1}^{p} B_k(s + t_{N,n})\ell_{M,m}(t_{N,n} - \tau_k)$$

$$+ \int_{-\tau_{k(t_{N,n})+1}}^{-t_{N,n}} C_{k(t_{N,n})+1}(s + t_{N,n}, \theta)\ell_{M,m}(t_{N,n} + \theta)\,d\theta$$

$$+ \sum_{k=k(t_{N,n})+2}^{p} \int_{-\tau_k}^{-\tau_{k-1}} C_k(s + t_{N,n}, \theta)\ell_{M,m}(t_{N,n} + \theta)\,d\theta,$$

for $n = 1, \ldots, \hat{N}$ and $m = 0, 1, \ldots, M$. This matrix is constructed in the lines 111–231. Integrals and Lagrange coefficients are computed as explained in Sect. 7.3.3.

If $h < \tau$, P_M in (7.31) is based on the piecewise mesh Ω_M (7.26). We have to establish whether $t_{N,n} - \tau_k$ and $t_{N,n} + \theta$ fall in $[0, h]$, where $V_1 P_M \Phi$ has the constant value $\Phi_0^{(1)}$, or in $[-qh, -(q-1)h]$, for $q = 1, \ldots, Q$, where $V_1 P_M \Phi$ coincides with the interpolating polynomial based on the nodes of $\Omega_M^{(q)}$ in (7.27), for $q = 1, \ldots, Q - 1$ or those of $\Omega_M^{(Q)}$ in (7.28) for $q = Q$. Let

$$\hat{N} := \begin{cases} 0 & \text{if } t_{N,n} \geq \tau - (Q-1)h \text{ for all } n = 1, \ldots, N, \\ \max\{n : t_{N,n} < \tau - (Q-1)h\} & \text{otherwise}, \end{cases}$$

$$q_n := \begin{cases} Q & \text{if } n = 1, \ldots, \hat{N}, \\ Q - 1 & \text{if } n = \hat{N} + 1, \ldots, N, \end{cases}$$

and

$$t_{N,n}^{(q)} := \begin{cases} t_{N,n} + qh & \text{if } q = 0, 1, \ldots, q_n - 1, \\ \tau & \text{if } q = q_n. \end{cases}$$

7.3 The Solution Operator Approach

Moreover, we still use (7.38). With a little patience, following the case $h \geq \tau$, it is tedious but not difficult to obtain the resulting matrix in $\mathbb{R}^{dN \times d(QM+1)}$

$$U_{M,N}^{(1)} = \begin{pmatrix} F_{1,0}^{(1)} & \cdots & F_{1,M-1}^{(1)} & F_{1,0}^{(2)} & \cdots & F_{1,0}^{(Q)} & F_{1,1}^{(Q)} & \cdots & F_{1,M-1}^{(Q)} & F_{1,M}^{(Q)} \\ \vdots & & \vdots & \vdots & & \vdots & \vdots & & \vdots & \vdots \\ F_{\hat{N},0}^{(1)} & \cdots & F_{\hat{N},M-1}^{(1)} & F_{\hat{N},0}^{(2)} & \cdots & F_{\hat{N},0}^{(Q)} & F_{\hat{N},1}^{(Q)} & \cdots & F_{\hat{N},M-1}^{(Q)} & F_{\hat{N},M}^{(Q)} \\ F_{\hat{N}+1,0}^{(1)} & \cdots & F_{\hat{N}+1,M-1}^{(1)} & F_{\hat{N}+1,0}^{(2)} & \cdots & F_{\hat{N}+1,0}^{(Q)} & 0 & \cdots & 0 & 0 \\ \vdots & & \vdots & \vdots & & \vdots & \vdots & & \vdots & \vdots \\ F_{N,0}^{(1)} & \cdots & F_{N,M-1}^{(1)} & F_{N,0}^{(2)} & \cdots & F_{N,0}^{(Q)} & 0 & \cdots & 0 & 0 \end{pmatrix},$$

where

$$F_{n,0}^{(1)} = A(s + t_{N,n}) + \sum_{k=1}^{k(t_{N,n})} B_k(s + t_{N,n}) + \sum_{k=k(t_{N,n})+1}^{k(t_{N,n}^{(1)})} B_k(s + t_{N,n}) \ell_{M,0}^{(1)}(t_{N,n} - \tau_k)$$

$$+ \sum_{k=1}^{k(t_{N,n})} \int_{-\tau_k}^{-\tau_{k-1}} C_k(s + t_{N,n}, \theta) \, d\theta + \int_{-t_{N,n}}^{-\tau_{k(t_{N,n})}} C_{k(t_{N,n})+1}(s + t_{N,n}, \theta) \, d\theta$$

$$+ \int_{-t_{N,n}^{(1)}}^{-\tau_{k(t_{N,n}^{(1)})}} C_{k(t_{N,n}^{(1)})+1}(s + t_{N,n}, \theta) \ell_{M,0}^{(1)}(t_{N,n} + \theta) \, d\theta$$

$$+ \sum_{k=k(t_{N,n})+2}^{k(t_{N,n}^{(1)})} \int_{-\tau_k}^{-\tau_{k-1}} C_k(s + t_{N,n}, \theta) \ell_{M,0}^{(1)}(t_{N,n} + \theta) \, d\theta$$

$$+ \int_{-\tau_{k(t_{N,n})+1}}^{-t_{N,n}} C_{k(t_{N,n})+1}(s + t_{N,n}, \theta) \ell_{M,0}^{(1)}(t_{N,n} + \theta) \, d\theta$$

for $n = 1, \ldots, N$,

$$F_{n,0}^{(q)} = \sum_{k=k(t_{N,n}^{(q-2)})+1}^{k(t_{N,n}^{(q-1)})} B_k(s + t_{N,n}) \ell_{M,M}^{(q-1)}(t_{N,n} - \tau_k)$$

$$+ \sum_{k=k(t_{N,n}^{(q-1)})+1}^{k(t_{N,n}^{(q)})} B_k(s + t_{N,n}) \ell_{M,0}^{(q)}(t_{N,n} - \tau_k)$$

$$+ \int_{-t_{N,n}^{(q-1)}}^{-\tau_{k(t_{N,n}^{(q-1)})}} C_{k(t_{N,n}^{(q-1)})+1}(s+t_{N,n},\theta)\ell_{M,M}^{(q-1)}(t_{N,n}+\theta)\,d\theta$$

$$+ \sum_{k=k(t_{N,n}^{(q-2)})+2}^{k(t_{N,n}^{(q-1)})} \int_{-\tau_k}^{-\tau_{k-1}} C_k(s+t_{N,n},\theta)\ell_{M,M}^{(q-1)}(t_{N,n}+\theta)\,d\theta$$

$$+ \int_{-\tau_{k(t_{N,n}^{(q-2)})+1}}^{-t_{N,n}^{(q-2)}} C_{k(t_{N,n}^{(q-2)})+1}(s+t_{N,n},\theta)\ell_{M,M}^{(q-1)}(t_{N,n}+\theta)\,d\theta$$

$$+ \int_{-t_{N,n}^{(q)}}^{-\tau_{k(t_{N,n}^{(q)})}} C_{k(t_{N,n}^{(q)})+1}(s+t_{N,n},\theta)\ell_{M,0}^{(q)}(t_{N,n}+\theta)\,d\theta$$

$$+ \sum_{k=k(t_{N,n}^{(q-1)})+2}^{k(t_{N,n}^{(q)})} \int_{-\tau_k}^{-\tau_{k-1}} C_k(s+t_{N,n},\theta)\ell_{M,0}^{(q)}(t_{N,n}+\theta)\,d\theta$$

$$+ \int_{-\tau_{k(t_{N,n}^{(q-1)})+1}}^{-t_{N,n}^{(q-1)}} C_{k(t_{N,n}^{(q-1)})+1}(s+t_{N,n},\theta)\ell_{M,0}^{(q)}(t_{N,n}+\theta)\,d\theta$$

for $n = 1, \ldots, N$ and $q = 2, \ldots, q_n$,

$$F_{n,m}^{(q)} = \sum_{k=k(t_{N,n}^{(q-1)})+1}^{k(t_{N,n}^{(q)})} B_k(s+t_{N,n})\ell_{M,m}^{(q)}(t_{N,n}-\tau_k)$$

$$+ \int_{-t_{N,n}^{(q)}}^{-\tau_{k(t_{N,n}^{(q)})}} C_{k(t_{N,n}^{(q)})+1}(s+t_{N,n},\theta)\ell_{M,m}^{(q)}(t_{N,n}+\theta)\,d\theta$$

$$+ \sum_{k=k(t_{N,n}^{(q-1)})+2}^{k(t_{N,n}^{(q)})} \int_{-\tau_k}^{-\tau_{k-1}} C_k(s+t_{N,n},\theta)\ell_{M,m}^{(q)}(t_{N,n}+\theta)\,d\theta$$

7.3 The Solution Operator Approach

$$+ \int_{-\tau_{k(t_{N,n}^{(q-1)})+1}}^{-t_{N,n}^{(q-1)}} C_{k(t_{N,n}^{(q-1)})+1}(s+t_{N,n},\theta) \ell_{M,m}^{(q)}(t_{N,n}+\theta)\,d\theta,$$

for $n = 1, \ldots, N$, $q = 1, \ldots, q_n$ and $m = 1, \ldots, M-1$,

$$F_{n,M}^{(Q)} = \sum_{k=k(t_{N,n}^{(Q-1)})+1}^{p} B_k(s+t_{N,n}) \ell_{M,M}^{(Q)}(t_{N,n} - \tau_k)$$

$$+ \sum_{k=k(t_{N,n}^{(Q-1)})+2}^{p} \int_{-\tau_k}^{-\tau_{k-1}} C_k(s+t_{N,n},\theta) \ell_{M,M}^{(Q)}(t_{N,n}+\theta)\,d\theta$$

$$+ \int_{-\tau_{k(t_{N,n}^{(Q-1)})+1}}^{-t_{N,n}^{(Q-1)}} C_{k(t_{N,n}^{(Q-1)})+1}(s+t_{N,n},\theta) \ell_{M,M}^{(Q)}(t_{N,n}+\theta)\,d\theta,$$

for $n = 1, \ldots, \hat{N}$ and

$$F_{n,0}^{(Q)} = \sum_{k=k(t_{N,n}^{(Q-2)})+1}^{p} B_k(s+t_{N,n}) \ell_{M,M}^{(Q-1)}(t_{N,n} - \tau_k)$$

$$+ \sum_{k=k(t_{N,n}^{(Q-2)})+2}^{p} \int_{-\tau_k}^{-\tau_{k-1}} C_k(s+t_{N,n},\theta) \ell_{M,M}^{(Q-1)}(t_{N,n}+\theta)\,d\theta$$

$$+ \int_{-\tau_{k(t_{N,n}^{(Q-2)})+1}}^{-t_{N,n}^{(Q-2)}} C_{k(t_{N,n}^{(Q-2)})+1}(s+t_{N,n},\theta) \ell_{M,M}^{(Q-1)}(t_{N,n}+\theta)\,d\theta,$$

for $n = \hat{N}+1, \ldots, N$. This matrix is constructed in the lines 233–808. Integrals and Lagrange coefficients are computed as usual.

7.3.5 The Matrix $U_N^{(2)}$

From (7.18) and (7.19) we have, according to the mesh Ω_N^+ in (7.30),

$$[U_N^{(2)} Z]_n = A(s+t_{N,n})(V_2 P_N^+ Z)(t_{N,n}) + \sum_{k=1}^{p} B_k(s+t_{N,n})(V_2 P_N^+ Z)(t_{N,n} - \tau_k)$$

$$+ \sum_{k=1}^{p} \int_{-\tau_k}^{-\tau_{k-1}} C_k(s + t_{N,n}, \theta)(V_2 P_N^+ Z)(t_{N,n} + \theta) \, d\theta,$$

for $n = 1, \ldots, N$. P_N^+ in (7.36) is based on Ω_N^+. We have to establish whether $t_{N,n} - \tau_k$ and $t_{N,n} + \theta$ fall in $[0, h]$, where $V_2 P_N^+ Z$ coincides with the integral of $P_N^+ Z$, or in $[-\tau, 0]$, where $V_2 P_N^+ Z$ is zero. By (7.37) and (7.38), it is not difficult to get

$$[U_N^{(2)} Z]_n = A(s + t_{N,n}) \int_0^{t_{N,n}} \sum_{i=1}^{n} \ell_{N,i}^+(t) Z_i \, dt + \sum_{k=1}^{k^*} B_k(s + t_{N,n}) \int_0^{t_{N,n} - \tau_k} \sum_{i=1}^{n} \ell_{N,i}^+(t) Z_i \, dt$$
$$+ \sum_{k=1}^{k^*} \int_{-\tau_k}^{-\tau_{k-1}} C_k(s + t_{N,n}, \theta) \left(\int_0^{t_{N,n} + \theta} \sum_{i=1}^{n} \ell_{N,i}^+(t) Z_i \, dt \right) d\theta$$
$$+ \int_{-\min\{\tau, t_{N,n}\}}^{-\tau_{k^*}} C_{k^*+1}(s + t_{N,n}, \theta) \left(\int_0^{t_{N,n} + \theta} \sum_{i=1}^{n} \ell_{N,i}^+(t) Z_i \, dt \right) d\theta,$$

for $n = 1, \ldots, N$ and $k^* := k(\min\{\tau, t_{N,n}\})$. The resulting matrix in $\mathbb{R}^{dn \times dN}$ reads

$$U_N^{(2)} = \begin{pmatrix} U_{1,1} & \cdots & U_{1,N} \\ \vdots & \ddots & \vdots \\ U_{N,1} & \cdots & U_{N,N} \end{pmatrix},$$

with

$$U_{n,i} = A(s + t_{N,n}) \int_0^{t_{N,n}} \ell_{N,i}^+(t) \, dt + \sum_{k=1}^{k^*} B_k(s + t_{N,n}) \int_0^{t_{N,n} - \tau_k} \ell_{N,i}^+(t) \, dt$$
$$+ \sum_{k=1}^{k^*} \int_{-\tau_k}^{-\tau_{k-1}} C_k(s + t_{N,n}, \theta) \left(\int_0^{t_{N,n} + \theta} \ell_{N,i}^+(t) \, dt \right) d\theta$$
$$+ \int_{-\min\{\tau, t_{N,n}\}}^{-\tau_{k^*}} C_{k^*+1}(s + t_{N,n}, \theta) \left(\int_0^{t_{N,n} + \theta} \ell_{N,i}^+(t) \, dt \right) d\theta,$$

for $n, i = 1, \ldots, N$. This matrix is constructed in the lines 811–900. Integrals and Lagrange coefficients are computed as usual. Note, however, the complication due to the presence of double integrals.

Chapter 8
Applications

All the following tests and applications refer to the notation and structure of model (7.2), i.e., from the user's point of view as explained in Sect. 7.1.

8.1 Test Cases

In this section, we analyze a series of (mostly academic) case studies in order to test the features and performance of eigAM.m and eigTMN.m.

8.1.1 Test 1: Linear Autonomous Equations with a Discrete Delay

Consider the Hayes equation (1.4) rewritten according to (7.2), i.e.,

$$x'(t) = \tilde{a}x(t) + \tilde{b}x(t-1), \qquad (8.1)$$

with $\tilde{a}, \tilde{b} \in \mathbb{R}$. The delay is $d_1 = 1$ without loss of generality. A simplified version of the stability chart shown in Fig. 1.1 is represented in Fig. 8.1 (left). Gray corresponds to the choices of the parameters $\tilde{a}$ and $\tilde{b}$ for which the zero solution of (8.1) is asymptotically stable, white where it is unstable. Along the thick solid lines, the rightmost characteristic root is on the imaginary axis: a real root at 0 along the upper line $\tilde{a} + \tilde{b} = 0$ (fold bifurcation, [127]) and an imaginary couple $\pm i\beta$, $\beta > 0$, along the lower line (Hopf bifurcation, [127]). Thin solid lines correspond to successive crossings of the imaginary axis. Along the dashed line, a real root with double multiplicity is present. As already explained in Sect. 1.2, the equations of all such curves can be found, e.g., in [33, 106], and the number of roots with positive real part in each white portion can be determined by the above considerations.

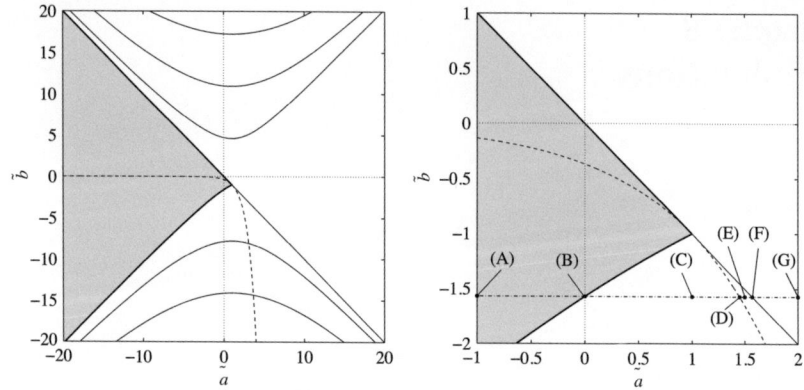

Fig. 8.1 Stability chart of (8.1) (*left*, *gray* region is stable, *white* unstable) and zoom (*right*, *dashed-dotted line* is for $\tilde{b} = -\pi/2$ and $\tilde{a} = -1$ (A), 0 (B), 1 (C), $1 + \log(\pi/2)$ (D), 1.5 (E), $\pi/2$ (F) and 2 (G))

We now fix $\tilde{b} = -\pi/2$ and thus perform some tests on the DDE

$$x'(t) = \tilde{a}x(t) - \frac{\pi}{2}x(t-1). \tag{8.2}$$

In particular, we let $\tilde{a}$ vary along the horizontal dashed-dotted line in the zoomed chart in Fig. 8.1 (right). The exact analysis of the characteristic equation reveals the following situation for the rightmost roots:

(A) $\tilde{a} = -1$: complex-conjugate pair with negative real part;
(B) $\tilde{a} = 0$: imaginary pair $\lambda = \pm i\pi/2$ (Hopf bifurcation);
(C) $\tilde{a} = 1$: complex-conjugate pair with positive real part;
(D) $\tilde{a} = 1 + \log(\pi/2)$: double real root $\lambda = \log(\pi/2)$;
(E) $\tilde{a} = 1.5$: two real roots, both positive;
(F) $\tilde{a} = \pi/2$: two real roots, one 0 and one positive;
(G) $\tilde{a} = 2$: two real roots, one negative and one positive.

These rightmost roots for varying $\tilde{a}$ are shown in Fig. 8.2. They are computed with eigAM.m, by calling, e.g., for $\tilde{a} = -1$, case (A),

```
>> [lambda,M]=eigAM('myDDE_test1',[-1,-pi/2],20);
```

with $M = 20$. The content of the script myDDE_test1.m relevant to (8.2) follows:

```
%% MEMO LIST OF POSSIBLE PARAMETERS
%par(1)=atilde;
%par(2)=btilde;

%% DIMENSION OF THE DDE
d=1;
```

8.1 Test Cases

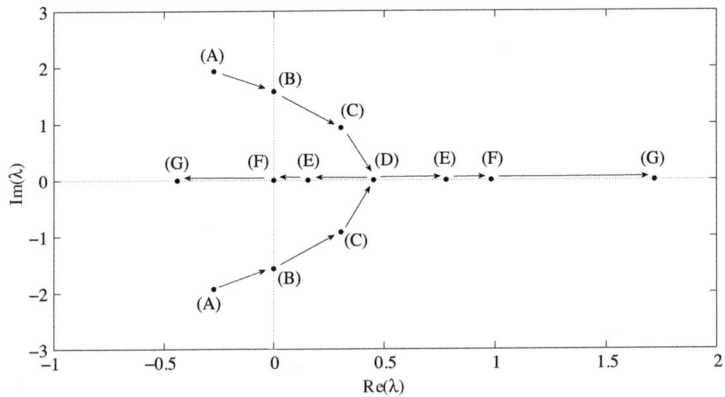

Fig. 8.2 Rightmost roots of (8.2) computed by eigAM.m with $M = 20$ for $\tilde{a} = -1$ (A), 0 (B), 1 (C), $1 + \log(\pi/2)$ (D), 1.5 (E), $\pi/2$ (F) and 2 (G) (*arrows* indicate increasing $\tilde{a}$)

```
%% CURRENT TIME TERM
Atilde=@(t,d,par) par(1);

%% DISCRETE DELAY TERMS
dd=1;
Btilde{1}=@(t,d,par) par(2);

%% DISTRIBUTED DELAY TERMS
l=[];
r=[];
Ctilde{1}=@(t,theta,d,par) [];
```

The above call returns the first $M + 1 = 21$ approximated roots, where the rightmost couple is, according to the theory, a complex-conjugate pair with negative real part:

```
>> lambda
lambda =
  -0.2728 + 1.9310i
  -0.2728 - 1.9310i
  -1.6025 + 7.7767i
  -1.6025 - 7.7767i
  -2.1948 +14.0523i
  -2.1948 -14.0523i
  -2.5640 +20.3435i
  -2.5640 -20.3435i
  -2.8350 +26.6490i
  -2.8350 -26.6490i
  -3.1929 +32.6281i
```

Table 8.1 Rightmost roots of (8.2) computed by eigAM.m with $M = 20$ for varying $\tilde{a}$

Case	$\tilde{a}$	λ_1	λ_2
(A)	-1	$-0.2728 + i1.9310$	$-0.2728 - i1.9310$
(B)	0	$-0.0000 + i1.5708$	$-0.0000 - i1.5708$
(C)	1	$0.3041 + i0.9267$	$0.3041 - i0.9267$
(D)	$1 + \log(\pi/2)$	0.4516	0.4516
(E)	1.5	0.7798	0.1557
(F)	$\pi/2$	0.9831	0.0000
(G)	2	1.7182	-0.4407

```
 -3.1929  -32.6281i
 -3.5176  +38.9579i
 -3.5176  -38.9579i
-14.2453  +43.5994i
-14.2453  -43.5994i
-17.1405  +70.8072i
-17.1405  -70.8072i
-49.2062  +43.2639i
-49.2062  -43.2639i
-74.4568
```

Eventually, Table 8.1 collects the two rightmost roots similarly obtained for varying $\tilde{a}$, confirming the theoretical findings previously mentioned.

We now analyze the convergence behavior of eigAM.m. We first consider (8.2) for $\tilde{a} = -1$, case (A). The first five rightmost complex-conjugate roots are shown in Fig. 8.3 (left), computed with $M = 20$, together with their error for increasing M

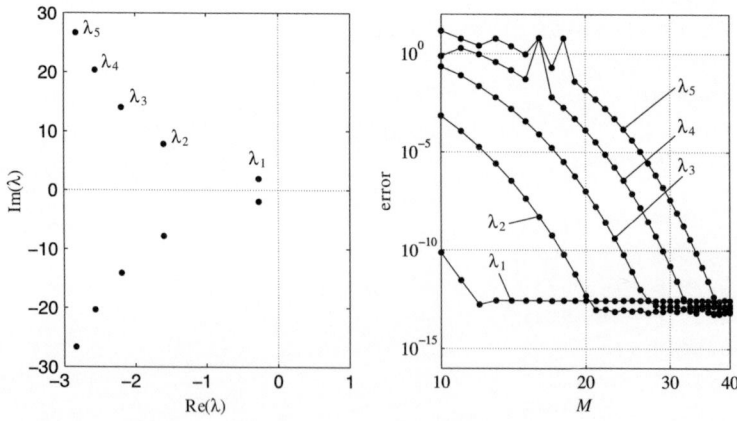

Fig. 8.3 The first five rightmost complex-conjugate roots of (8.2) computed by eigAM.m with $M = 20$ for $\tilde{a} = -1$, case (A) in Fig. 8.1, (*left*) and the corresponding error for increasing M (*right*)

8.1 Test Cases

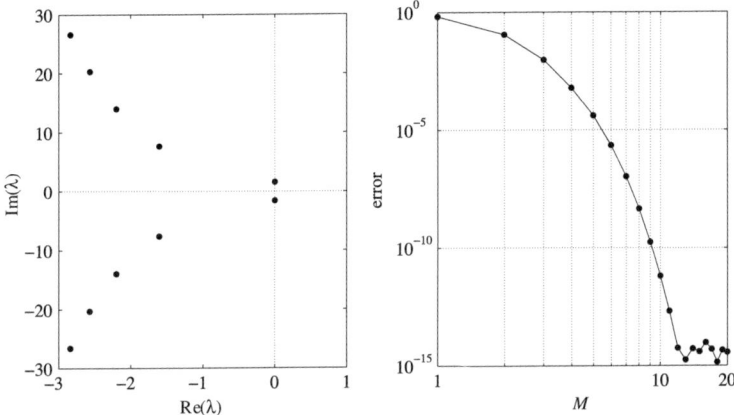

Fig. 8.4 The first five rightmost complex-conjugate roots of (8.2) computed by eigAM.m with $M = 20$ for $\tilde{a} = 0$, case (B) in Fig. 8.1, (*left*) and the error for increasing M for the rightmost couple $\lambda = \pm i\pi/2$ (*right*)

(right). For $\tilde{a} = -1$, there is no notion of the exact value of the roots. Then, as a standard procedure in testing numerical methods, we consider as exact those roots computed by eigAM.m for very large M, say $M = 200$. Note how the error of the considered roots decreases down to machine precision according to spectral accuracy, as stated in Theorem 5.2. Moreover, roots closer to the origin are approximated first, due to the presence of the constant C_1 in the error bound (5.30). In fact, as proved in Proposition 5.1, this constant is proportional to $|\lambda|$.

In Fig. 8.4, we report the same analysis for $\tilde{a} = 0$, case (B), the error being restricted to the rightmost complex-conjugate pair. In this case, the rightmost couple is exactly $\lambda = \pm i\pi/2$: the true error in the right panel confirms again Theorem 5.1.

The error behavior is the same for all the other cases (C-G). A further interesting aspect concerns case (D). In fact, for $\tilde{a} = 1 + \log(\pi/2)$, the theory ensures that the rightmost root is a double real root. This is already confirmed through eigAM.m in Fig. 8.2 and Table 8.1. As for convergence, Fig. 8.5 shows correctly that only half the machine precision can be reached. This is the effect of the double multiplicity at the denominator of the exponent in the error bound (5.31).

As a last test for (8.2), we consider the use of eigTMN.m for the approximation of the characteristic multipliers. As already remarked, eigTMN.m is devoted to nonautonomous (and, in particular, periodic) problems, of which autonomous ones are a special instance. If λ is a characteristic root, the corresponding multiplier is $\mu = e^{\lambda \tau}$, Proposition 4.4. Therefore, for, e.g., case (B), the rightmost complex-conjugate pair $\lambda = \pm i\pi/2$ corresponds to the dominant multipliers $\mu = \pm i$. Figure 8.6 (left) shows the spectrum of multipliers approximated by eigTMN.m with $M = N = 20$. The first two are indeed

```
>> mu=eigTMN('myDDE_test1',[0,-pi/2],0,1,20,20);
>> mu(1:2)
```

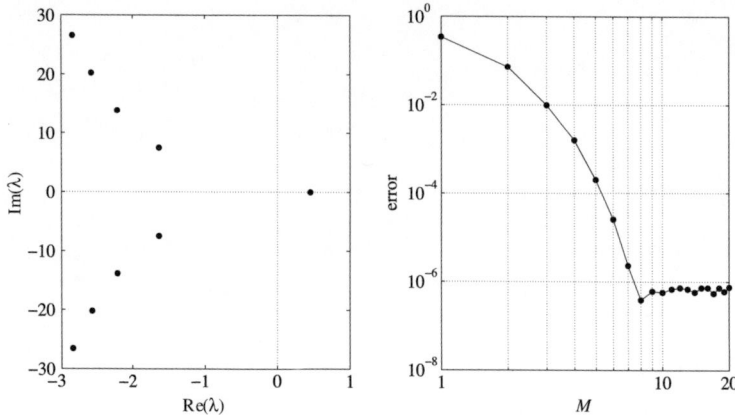

Fig. 8.5 The first rightmost roots of (8.2) computed by `eigAM.m` with $M = 20$ for $\tilde{a} = 1 + \log(\pi/2)$, case (D) in Fig. 8.1, (*left*) and the error for increasing M for the rightmost double root (*right*)

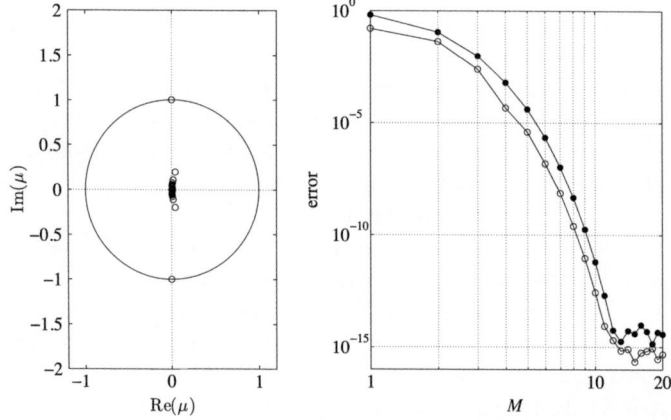

Fig. 8.6 Dominant multipliers of (8.2) computed by `eigTMN.m` with $M = N = 20$ for $\tilde{a} = 0$, case (B) in Fig. 8.1, (*left*) and the error for increasing M (and $N = M$) for the dominant multiplier $\mu = i$ with `eigTMN.m` (*right*, ∘) and for the rightmost root $\lambda = i\pi/2$ with `eigAM.m` (*right*, •)

```
ans =
   0.0000 + 1.0000i
   0.0000 - 1.0000i
```

In Fig. 8.6 (right), the error is shown for increasing $M = N$. For convenience, also the error of the rightmost root with `eigAM.m` is reported: both methods converge with spectral accuracy according to Theorem 5.1 and 6.6. A deeper analysis of `eigTMN.m` is performed, however, in Sect. 8.1.5 for a nonautonomous periodic DDE, and in Sect. 8.1.6, where the nonlinear autonomous DDE (4.10) of Example 4.2 is considered when linearized at a specific known periodic solution.

8.1.2 Test 2: Linear Autonomous Equations with Multiple Discrete Delays

The second test equation is a DDE with two discrete delay terms:

$$x'(t) = \left(1 - \frac{1}{e}\right)x(t) + \frac{1}{2}x(t - d_1) + \frac{1}{2}ex(t - d_2). \tag{8.3}$$

The relevant content of the corresponding script myDDE_test2.m is

```
%% MEMO LIST OF POSSIBLE PARAMETERS
%par(1)=d1;
%par(2)=d2;

%% DIMENSION OF THE DDE
d=1;

%% CURRENT TIME TERM
Atilde=@(t,d,par) 1-exp(-1);

%% DISCRETE DELAY TERMS
dd=[par(1),par(2)];
Btilde{1}=@(t,d,par) .5;
Btilde{2}=@(t,d,par) .5*exp(1);

%% DISTRIBUTED DELAY TERMS
l=[];
r=[];
Ctilde{1}=@(t,theta,d,par) [];
```

It is not difficult to verify that equation (8.3) has an exact real rightmost root $\lambda = 1$ for $d_1 = 1$ and $d_2 = 2$. The results of the approximation of the rightmost roots are presented in Fig. 8.7, in the same spirit of the previous section. Spectral accuracy is confirmed also in this case with multiple delays. Note, however, that having equation (8.3) two delays, the piecewise approach of Sect. 7.2.3 is adopted. Therefore, the M in the right figure refers to the total number of nodes actually used for the discretization (i.e., the one furnished in output by eigAM.m). The same spectral behavior persists even in the case of delays nonrationally dependent, as shown in Fig. 8.8 for $d_1 = 1$ and $d_2 = \sqrt{2}$. However, in this case, it is not possible to recover the exact value of the rightmost root, which is slightly above one:

```
>> [lambda,M]=eigAM('myDDE_test2',[1,sqrt(2)],20);
>> lambda(1)
ans =
    1.0907
```

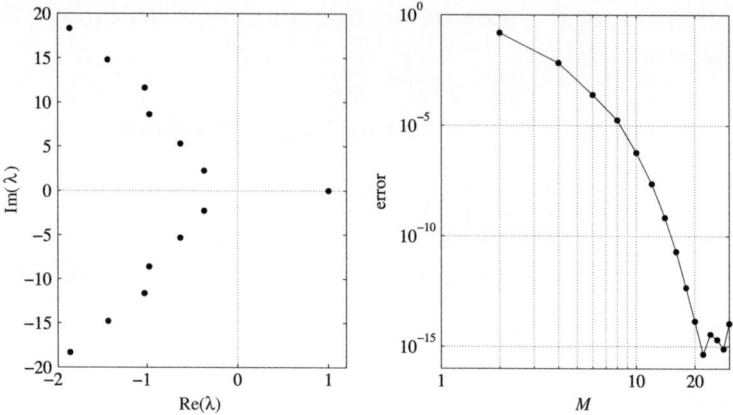

Fig. 8.7 The first rightmost roots of (8.3) computed by eigAM.m with $M = 20$ for $d_1 = 1$ and $d_2 = 2$ (*left*) and the error for increasing M for the rightmost root (*right*)

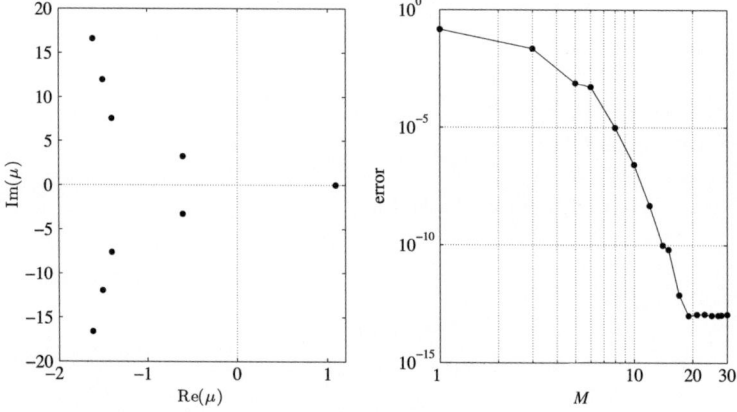

Fig. 8.8 The first rightmost roots of (8.3) computed by eigAM.m with $M = 20$ for $d_1 = 1$ and $d_2 = \sqrt{2}$ (*left*) and the error with eigAM.m for increasing M for the rightmost root (*right*)

8.1.3 Test 3: Linear Autonomous Equations with a Distributed Delay

In this section, we test eigAM.m on

$$x'(t) = \frac{1}{2}\left(1 + \frac{1}{e}\right)x(t) + \int_{-1/2}^{0} e^{\theta} x(t+\theta)\,\mathrm{d}\theta, \qquad (8.4)$$

a DDE with a single distributed delay term. The aim is at confirming the spectral convergence also in the case that quadrature is needed, see Sect. 7.2.2. As for (8.3),

8.1 Test Cases

it is not difficult to show that (8.4) has a real rightmost root $\lambda = 1$. The content of the associated script myDDE_test3.m is

```
%% MEMO LIST OF POSSIBLE PARAMETERS
%no parameters

%% DIMENSION OF THE DDE
d=1;

%% CURRENT TIME TERM
Atilde=@(t,d,par) .5*(1+exp(-1));

%% DISCRETE DELAY TERMS
dd=[];
Btilde{1}=@(t,d,par) [];

%% DISTRIBUTED DELAY TERMS
l=.5;
r=0;
Ctilde{1}=@(t,theta,d,par) exp(theta);
```

Figure 8.9 confirms again the spectral convergence for the rightmost root, which is correctly approximated:

```
>> [lambda,M]=eigAM('myDDE_test3',[],20);
>> lambda(1)
ans =
    1.0000
```

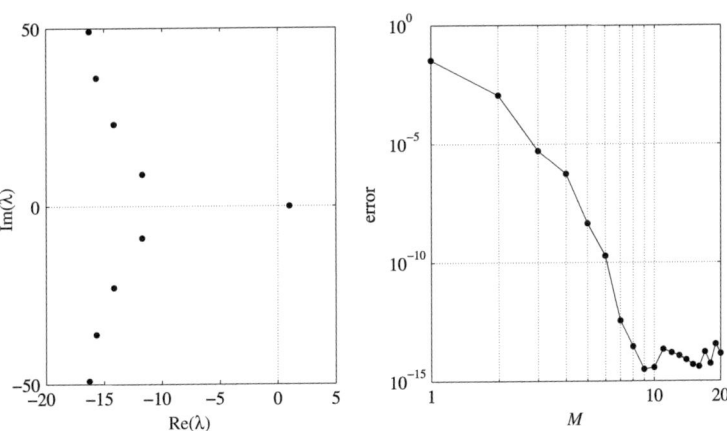

Fig. 8.9 The first rightmost roots of (8.4) computed by eigAM.m with $M = 20$ (*left*) and the error for increasing M for the rightmost root (*right*)

8.1.4 Test 4: Linear Autonomous Systems

As a last test for eigAM.m, we consider the system of two DDEs

$$\begin{cases} x_1'(t) = x_1(t) + ex_2(t) + x_2(t-1) + 2\int_{-1}^{0} \theta x_1(t+\theta)\,d\theta, \\ x_2'(t) = ex_1(t) - x_2(t) - x_1(t-1) + \int_{-1}^{0} x_2(t+\theta)\,d\theta, \end{cases} \quad (8.5)$$

both with discrete and distributed delay terms. The content of myDDE_test4.m is

```
%% MEMO LIST OF POSSIBLE PARAMETERS
%no parameters

%% DIMENSION OF THE DDE
d=2;

%% CURRENT TIME TERM
Atilde=@(t,d,par) [1,exp(1);exp(1),-1];

%% DISCRETE DELAY TERMS
dd=1;
Btilde{1}=@(t,d,par) [0,1;-1,0];

%% DISTRIBUTED DELAY TERMS
l=1;
r=0;
Ctilde{1}=@(t,theta,d,par) [2*theta,0;0,1];
```

System (8.5) has the exact real root $\lambda = -1$, and Fig. 8.10 confirms again the spectral convergence. This is not the rightmost one, in fact:

```
>> [lambda,M]=eigAM('myDDE_test4',[],20);
>> lambda(1:2)
ans =
    2.8869
   -1.0000
```

8.1 Test Cases

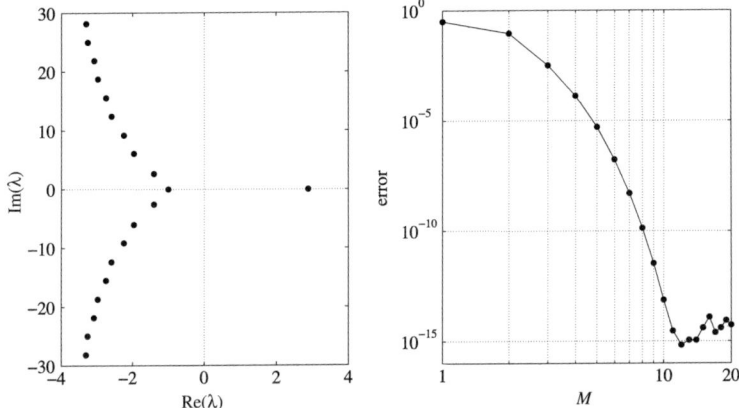

Fig. 8.10 The first rightmost roots of (8.5) computed by eigAM.m with $M = 20$ (*left*) and the error for increasing M for the real root $\lambda = -1$ (*right*)

8.1.5 Test 5: Linear Periodic Equations

At the end of Sect. 8.1.1, eigTMN.m is tested on the autonomous DDE (8.2). Here, instead, we consider the approximation of the characteristic multipliers for a DDE with nonconstant periodic coefficients. In particular, let $\rho, \sigma \in \mathbb{R}$ and consider

$$x'(t) = \tilde{a}(t)x(t) + \tilde{b}(t)x(t - 2\pi) + \int_{-2\pi}^{0} \tilde{c}(t, \theta) x(t + \theta) \, d\theta, \qquad (8.6)$$

where $\tilde{a}(t) = \cos(t)$, $\tilde{b}(t) = \sin(t) + \rho e^{2\pi(\rho+\sigma)}$ and

$$\tilde{c}(t, \theta) = \frac{\sigma^2 e^{\sin(t)}[1-\cos(\theta)-e^{-2\pi(\rho+\sigma)}\sin(\theta)]-\cos(t)[\sin(\theta)+e^{-2\pi(\rho+\sigma)}(1-\cos(\theta))]-\rho\theta}{1 - e^{-2\pi\sigma}}.$$

The content of the associated script myDDE_test5.m is

```
%% MEMO LIST OF POSSIBLE PARAMETERS
%par(1)=rho;
%par(2)=sigma;

%% DIMENSION OF THE DDE
d=1;

%% CURRENT TIME TERM
Atilde=@(t,d,par) cos(t);
```

```
%% DISCRETE DELAY TERMS
dd=2*pi;
Btilde{1}=@(t,d,par) sin(t)+par(1)*...
    exp(2*pi*(par(1)+par(2)));

%% DISTRIBUTED DELAY TERMS
l=2*pi;
r=0;
Ctilde{1}=@(t,theta,d,par) (par(2))^2*exp(sin(t)-...
    sin(t)*cos(theta)-cos(t)*sin(theta)+...
    exp(-2*pi*(par(1)+par(2)))*(cos(t)*cos(theta)-...
    sin(t)*sin(theta)-cos(t))-theta*par(1))/...
    (1-exp(-2*pi*par(2)));
```

For these 2π-periodic coefficients (8.6) has an exact multiplier $\mu = e^{2\pi(\rho+\sigma)}$ (it can be proved by the arguments used in [91, Sect. 8.1] for the DDE (1.5) in there).

For, e.g., $\rho = 1$ and $\sigma = -0.85$, it turns out that $\mu = 2.566332395208135$ is the dominant multiplier, correct to machine precision. Indeed eigTMN.m gives:

```
>> rho=1;sigma=-.85;
>> mu=eigTMN('myDDE_test5',[rho,sigma],0,2*pi,20,20);
>> mu(1)
ans =
    2.5663
```

Finally, Fig. 8.11 confirms again the spectral convergence also for the approximation of the multipliers by eigTMN.m. Note that (8.6) has also a distributed delay term, which is correctly approximated by the quadrature procedure.

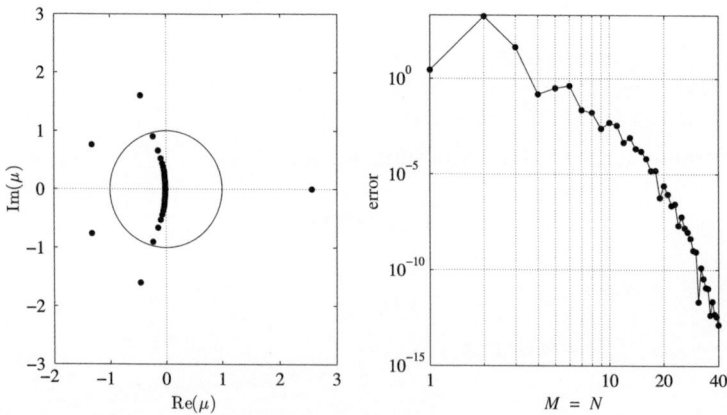

Fig. 8.11 The first dominant multipliers of (8.6) for $\rho = 1$ and $\sigma = -0.85$ computed by eigTMN.m with $M = N = 40$ (*left*) and the error for increasing M (and $N = M$) for the dominant multiplier $\mu = e^{2\pi(\rho+\sigma)} \simeq 2.5663$ (*right*)

8.1.6 Test 6: Linearized Periodic Equations

To further test eigTMN.m, we consider the linear periodic DDE (4.11), i.e.,

$$x'(t) = \cos(t)x(t) - e^{\sin(t)+\cos(t)}x\left(t - \frac{\pi}{2}\right), \tag{8.7}$$

obtained from the linearization of the nonlinear autonomous DDE (4.10) at the exact 2π-periodic solution $\bar{x}(t) = e^{\sin(t)}$. The script myDDE_test6.m contains

```
%% MEMO LIST OF POSSIBLE PARAMETERS
%no parameters

%% DIMENSION OF THE DDE
d=1;

%% CURRENT TIME TERM
Atilde=@(t,d,par) cos(t);

%% DISCRETE DELAY TERMS
dd=pi/2;
Btilde{1}=@(t,d,par) -exp(sin(t)+cos(t));

%% DISTRIBUTED DELAY TERMS
l=[];
r=[];
Ctilde{1}=@(t,theta,d,par) [];
```

and the test is devoted to verify numerically the presence of the characteristic multiplier $\mu = 1$, due to the linearization procedure as explained in Sect. 4.2. Indeed:

```
>> mu=eigTMN('myDDE_test6',[],0,2*pi,40,40);
>> mu(1)
ans =
    1.0000
```

and Fig. 8.12 (right) confirms again the spectral convergence. Note also from Fig. 8.12 (left) that all the other computed multipliers are inside the unit circle: the periodic solution $\bar{x}$ is exponentially asymptotically stable by Theorem 4.3.

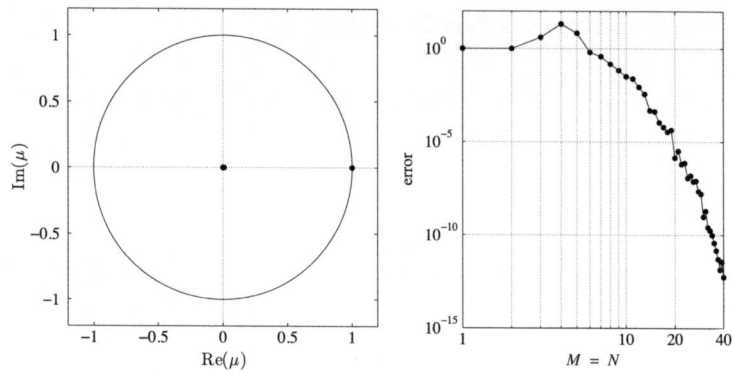

Fig. 8.12 The first dominant multipliers of (8.7) computed by `eigTMN.m` with $M = N = 40$ (*left*) and the error for increasing M (and $N = M$) for the dominant multiplier $\mu = 1$ (*right*)

8.2 Equilibria in Population Dynamics

In this section, we apply `eigAM.m` to analyze the asymptotic stability of the equilibria of the predator–prey model of Beddington-DeAngelis type with maturation and gestation delays studied in [133]. Denoting by x_1, x_2 and x_3 the densities of, respectively, preys and adult and juvenile predators, the model reads

$$\begin{cases} x_1'(t) = rx_1(t)\left(1 - \dfrac{x_1(t)}{K}\right) - \dfrac{bx_1(t)x_2(t)}{1 + k_1 x_1(t) + k_2 x_2(t)}, \\[1em] x_2'(t) = \dfrac{nbe^{-m_j s} x_1(t-s-s_1) x_2(t-s)}{1 + k_1 x_1(t-s-s_1) + k_2 x_2(t-s-s_1)} - mx_2(t), \\[1em] x_3'(t) = \dfrac{nbx_1(t-s_1) x_2(t)}{1 + k_1 x_1(t-s_1) + k_2 x_2(t-s_1)} - \dfrac{nbe^{-m_j s} x_1(t-s-s_1) x_2(t-s)}{1 + k_1 x_1(t-s-s_1) + k_2 x_2(t-s-s_1)} \\[1em] \qquad\quad - m_j x_3(t) \end{cases}$$

according to the following assumptions:

(A1) in the absence of predators, preys are subject to a logistic growth with maximum growth rate r and carrying capacity K [135, 157, 192, 193];
(A2) the per-capita predation rate is the Beddington-DeAngelis function with maximum predation rate b, handling time k_1 and interference k_2 [7, 10, 17, 56, 65, 68, 99, 100, 175, 199];
(A3) the predator population is divided into juveniles and adults and the former enter the latter stage after a maturation delay s [132];
(A4) juvenile predators dye at a specific mortality rate m_j while the specific mortality rate of the adults is m [132];

8.2 Equilibria in Population Dynamics

(A5) the juvenile predators growth rate at time t is proportional to the density of adult predators at the same time t times the amount of preys captured at some previous time $t - s_1$ (s_1 being a gestation delay [58, 134, 168]) through a biomass transformation n [133];

(A6) all parameters are positive.

The notation is slightly modified w.r.t. [133] to avoid later confusion w.r.t. (7.2).

Since the equation for x_3 is a linear ODE with forcing term depending only on x_1 and x_2, it follows that the dynamics of the model is completely determined by the first two equations. Thus, we reduce to study the system of a nonlinear ODE coupled with a nonlinear DDE with two discrete delay terms

$$\begin{cases} x_1'(t) = rx_1(t)\left(1 - \dfrac{x_1(t)}{K}\right) - \dfrac{bx_1(t)x_2(t)}{1 + k_1x_1(t) + k_2x_2(t)}, \\ x_2'(t) = \dfrac{nbe^{-m_j\tau_1}x_1(t-d_2)x_2(t-d_1)}{1 + k_1x_1(t-d_2) + k_2x_2(t-d_2)} - dx_2(t), \end{cases} \quad (8.8)$$

for $d_1 := s > 0$ and $d_2 := s + s_1 > d_1$.

The first step in our analysis is to find the equilibria of (8.8), i.e., the couples $E = (\bar{x}_1, \bar{x}_2)$ for $\bar{x}_1$ and $\bar{x}_2$ nonnegative (biologically meaningful) solutions of

$$\begin{cases} rx_1\left(1 - \dfrac{x_1}{K}\right) - \dfrac{bx_1x_2}{1 + k_1x_1 + k_2x_2} = 0, \\ \dfrac{nbe^{-m_j\tau_1}x_1x_2}{1 + k_1x_1 + k_2x_2} - mx_2 = 0. \end{cases}$$

It is not difficult to obtain the three possible equilibria $E_0 = (0,0)$, $E_K = (K,0)$ and $E_+ = (x_1^+, x_2^+)$ with positive x_1^+ and x_2^+ given, respectively, by

$$\begin{cases} x_1^+ = \dfrac{1}{2}\left(-c_1 + \sqrt{c_1^2 + 4c_2}\right), \\ x_2^+ = \dfrac{x_1^+(nbe^{-m_j\tau_1} - mk_1) - m}{mk_2}, \end{cases} \quad (8.9)$$

for

$$c_1 := \dfrac{K}{r}\left(\dfrac{nbe^{-m_jd_1} - mk_1}{ne^{-m_jd_1}k_2} - r\right)$$

and

$$c_2 := \dfrac{Km}{rne^{-m_jd_1}k_2}.$$

Note that E_0 and E_K exist for all the values of the parameters while, as shown in [133], E_+ exists if and only if $d_1 \in (0, d^+)$ for

$$d^+ := \frac{1}{m_j} \log\left(\frac{Knb}{m(1+k_1K)}\right). \tag{8.10}$$

As a second step, we linearize (8.8) around the general equilibrium $E = (\bar{x}_1, \bar{x}_2)$ obtaining the 2-dimensional linear autonomous DDE

$$x'(t) = \tilde{A}x(t) + \tilde{B}_1 x(t-d_1) + \tilde{B}_2 x(t-d_2) \tag{8.11}$$

for $x(t) = (x_1(t), x_2(t))^T \in \mathbb{R}^2$, where it is not difficult to recover the matrices

$$\tilde{A} = \begin{pmatrix} r\left(1 - \frac{2\bar{x}_1}{K}\right) - \frac{b\bar{x}_2(1+k_2\bar{x}_2)}{(1+k_1\bar{x}_1+k_2\bar{x}_2)^2} & -\frac{b\bar{x}_1(1+k_1\bar{x}_1)}{(1+k_1\bar{x}_1+k_2\bar{x}_2)^2} \\ 0 & -m \end{pmatrix},$$

$$\tilde{B}_1 = \begin{pmatrix} 0 & 0 \\ 0 & \dfrac{ne^{-m_j d_1} b\bar{x}_1}{1+k_1\bar{x}_1+k_2\bar{x}_2} \end{pmatrix}$$

and

$$\tilde{B}_2 = \begin{pmatrix} 0 & 0 \\ \dfrac{ne^{-m_j d_1} b\bar{x}_2(1+k_2\bar{x}_2)}{(1+k_1\bar{x}_1+k_2\bar{x}_2)^2} & -\dfrac{ne^{-m_j d_1} k_2 b\bar{x}_1\bar{x}_2}{(1+k_1\bar{x}_1+k_2\bar{x}_2)^2} \end{pmatrix}.$$

As a final step to determine the stability of the equilibrium E, one would need to compute the roots of the characteristic equation

$$\det\left(\lambda I_2 - \tilde{A} - \tilde{B}_1 e^{-\lambda d_1} - \tilde{B}_2 e^{-\lambda d_2}\right) = 0.$$

The expression of the left-hand side is rather cumbersome, compare with [133]. There, the authors still managed to prove analytically that E_0 is an unstable saddle with $\lambda_1 = r > 0$ and $\lambda_2 = -m < 0$ and that E_K has always a characteristic root $\lambda = -r$. Concerning E_+, an exact analysis of the characteristic equation is not attainable anymore. We thus resort to eigAM.m to approximate the rightmost roots of (8.11). The same approach is used in [133] to inspect stability w.r.t. d_1 and d_2.

First, one has to write the relevant script myDDE_ppBDA.m:

```
%% MEMO LIST OF POSSIBLE PARAMETERS
%par(1)=r, preys growth rate
%par(2)=K, preys carrying capacity
```

8.2 Equilibria in Population Dynamics

```
%par(3)=b, predators capture rate
%par(4)=k1, predators handling time
%par(5)=k2, predators interference
%par(6)=n, predators birth rate
%par(7)=m, adult predators death rate
%par(8)=mj, juvenile predators death rate
%par(9)=s, predators maturation delay
%par(10=s1, predators gestation delay
%par(11)=xbar1, preys at equilibrium
%par(12)=xbar2, adult predators at equilibrium

%% DIMENSION OF THE DDE
d=2;

%% CURRENT TIME TERM
Atilde=@(t,d,par) [par(1)*(1--2*par(11)/...
    par(2))-par(3)*par(12)*(1+par(5)*...
    par(12))/((1+par(4)*par(11)+par(5)*...
    par(12))^2),-par(3)*par(11)*...
    (1+par(4)*par(11))/((1+par(4)*par(11)+...
    par(5)*par(12))^2);0,-par(7)];

%% DISCRETE DELAY TERMS
dd=[par(9),par(9)+par(10)];
Btilde{1}=@(t,d,par) [0,0;0,par(6)*...
    exp(-par(8)*par(9))*par(3)*par(11)/...
    (1+par(4)*par(11)+par(5)*par(12))];
Btilde{2}=@(t,d,par) [0,0;par(6)*...
    exp(-par(8)*par(9))*par(3)*par(12)*...
    (1+par(5)*par(12))/((1+par(4)*par(11)+...
    par(5)*par(12))^2),-par(6)*exp(-par(8)*...
    par(9))*par(5)*par(3)*par(11)*par(12)/...
    ((1+par(4)*par(11)+par(5)*par(12))^2)];

%% DISTRIBUTED DELAY TERMS
l=[];
r=[];
Ctilde{1}=@(t,theta,d,par) [];
```

Second, the parameters values are inserted. According, e.g., to (39) in [133], set

```
>> r=1;K=1.6;b=1.5;k1=1;k2=.1;n=1;m=.5;mj=.01;
>> s=1;s1=.1;
```

We perform an analysis for E_0, which we recall to have the exact roots $\lambda_1 = r = 1$ and $\lambda_2 = -m = -0.5$, independently of the delays. In fact, one obtains:

```
>> xbar1=0;xbar2=0;
>> [lambda,M]=eigAM('myDDE_ppBDA',...
[r,K,b,k1,k2,n,m,mj,s,s1,xbar1,xbar2],100);
>> lambda(1:2)
ans =
    1.0000
   -0.5000
```

As a second test, we can switch to the equilibrium E_K, which we recall it has the exact root $\lambda = -r = -1$, again independently of the delays:

```
>> xbar1=K;xbar2=0;
>> [lambda,M]=eigAM('myDDE_ppBDA',...
[r,K,b,k1,k2,n,m,mj,s,s1,xbar1,xbar2],100);
>> lambda(1:2)
ans =
    0.2278
   -1.0000
```

We can also plot the first rightmost roots to get an idea of the spectrum, Fig. 8.13.

As a further test, we turn to the positive equilibrium E_+. We first show that at the right boundary of its existence, i.e., for $d_1 = d^+$ in (8.10), the rightmost eigenvalue $\lambda = 0$ is expected due to a transcritical bifurcation [92] with E_K. In fact:

```
>> dplus=log(K*b*n/(m*(1+k1*K)))/mj
```

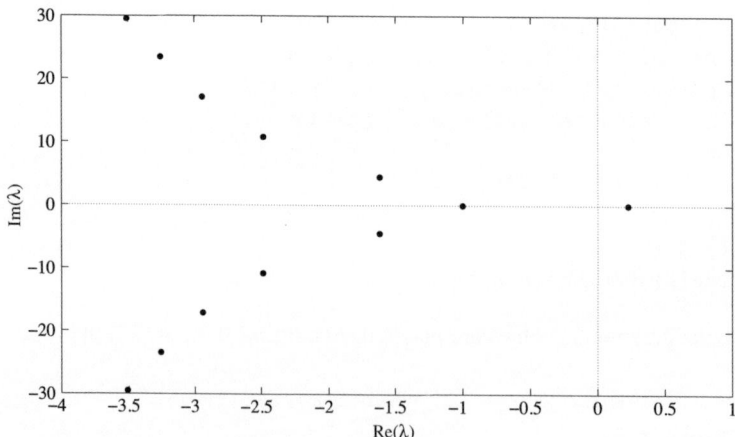

Fig. 8.13 The first rightmost roots of (8.11) for the equilibrium E_K computed by `eigAM.m` with $M = 100$

8.2 Equilibria in Population Dynamics

```
dplus =
   61.3104
>> s=dplus;s1=.1;
>> c1=K*((n*b*exp(-mj*s)-m*k1)/(n*exp(-mj*s)*k2)-r)/r;
>> c2=K*d/(n*exp(-mj*s)*k2*r);
>> xplus1=.5*(-c1+sqrt(c1^2+4*c2))
xplus1 =
   1.6000
>> xplus2=(xplus1*(n*b*exp(-mj*s)-j*k1)-m)/(m*k2)
xplus2 =
   0
>> [lambda,M]=eigAM('myDDE_ppBDA',...
[r,K,b,k1,k2,n,m,mj,s,s1,xplus1,xplus2],100);
>> lambda(1)
ans =
   -6.1427e-16
```

Part of the rightmost spectrum is shown in Fig. 8.14.

As a last example of the usefulness of eigAM.m, we show that (potential) Hopf bifurcations of the positive equilibrium E_+ are possible (as described in [133]), e.g.:

```
>> format long
>> s=50;s1=8.583284817754569;
>> c1=K*((n*b*exp(-mj*s)-m*k1)/(n*exp(-mj*s)*k2)-r)/r;
>> c2=K*m/(n*exp(-mj*s)*k2*r);
>> xplus1=.5*(-c1+sqrt(c1^2+4*c2))
```

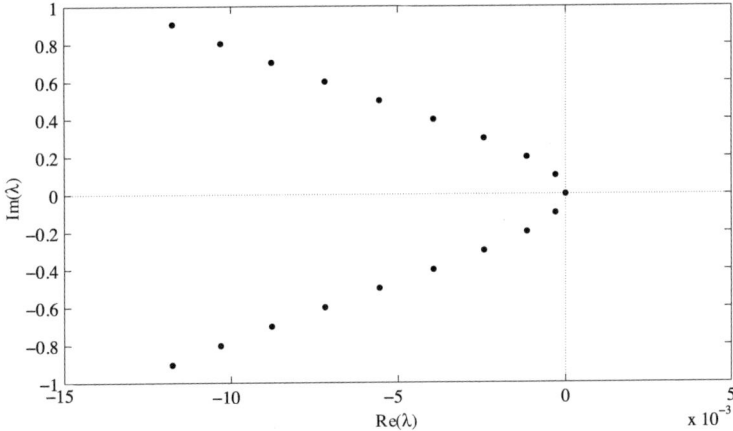

Fig. 8.14 The first rightmost roots of (8.11) for the equilibrium $E_+ = E_K$ at the transcritical bifurcation for $d_1 = d_+$ computed by eigAM.m with $M = 100$

```
xplus1 =
    1.259768164858865
>> xplus2=(xplus1*(n*b*exp(-mj*s)-m*k1)-m)/(m*k2)
xplus2 =
    0.324958834915974
>> [lambda,M]=eigAM('myDDE_ppBDA',...
[r,K,b,k1,k2,n,m,mj,s,s1,xplus1,xplus2],100);
>> lambda(1:2)
ans =
  -0.000000000000000 + 0.243999492780336i
  -0.000000000000000 - 0.243999492780336i
```

Part of the rightmost spectrum is shown in Fig. 8.15. The above bifurcation value for s_1 has been determined by the default root finder fzero.m of MATLAB:

```
>> fun=@(s1)max(real(eigAM('myDDE_ppBDA',...
[r,K,b,k1,k2,n,m,mj,s,s1,xplus1,xplus2],100)));
>> options=optimset('tolx',eps);
>> s1_bif=fzero(fun,0,options)
s1_bif =
    8.583284817754569
>> fun(s1_bif)
ans =
   -1.526556658859590e-16
```

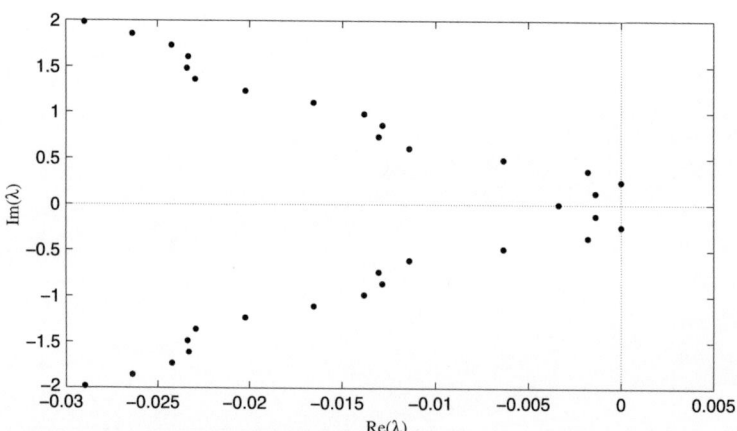

Fig. 8.15 The first rightmost roots of (8.11) for the equilibrium E_+ at a potential Hopf bifurcation for $d_1 = 50$ and $d_2 = 58.583284817753716$ computed by eigAM.m with $M = 100$

8.3 Periodic Problems in Engineering

In this section, we apply `eigTMN.m` to analyze the asymptotic stability of the zero solution of the delayed Mathieu equation

$$y''(t) + (\delta + \varepsilon \cos(t))y(t) = by(t - 2\pi). \tag{8.12}$$

This equation (and the variants considered later on) includes both the effect of the time delay and the action of parametric forcing appearing in several real-life applications from engineering. The periodicity is due to the presence of the parametric forcing. A first example, which is also recalled in the sequel, is given in Sect. 1.4.

For an extensive treatment of a large class of delayed Mathieu equations with regard to stability and applications, we refer again to the nice monograph [106]. See also the relevant literature through the references therein. In [106, Sect. 2.4], Insperger and Stepán studied the stability chart of (8.12) with reference to the variation of the three parameters δ, ε and b. The original [106, Fig. 2.10], reproduced here in Fig. 8.16

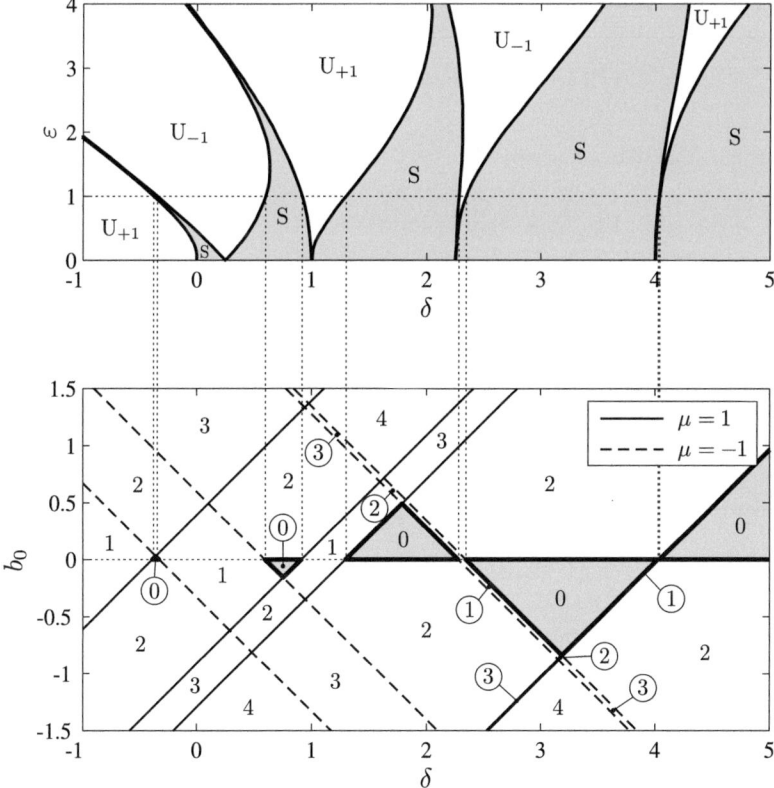

Fig. 8.16 Stability chart of the delayed Mathieu equation (8.12) for $\varepsilon = 1$ (*bottom*) with the number of unstable characteristic multipliers (b_0 is b in (8.12)). Original [106, Fig. 2.10] reproduced by courtesy of the authors and of Springer (copyright license 3387610410959, May 14, 2014)

by courtesy of the authors and of Springer (copyright license 3387610410959, May 14, 2014), was first published in [104], where the authors clarified the connection between the Strutt-Ince chart of the Mathieu equation (i.e., (8.12) with $b = 0$, see also [101, 158]) and the Hsu-Bhatt-Vyshnegradskii chart of the delayed oscillator (i.e., (8.12) with $\varepsilon = 0$, see also [97]). Here we first aim at approximating the dominant multipliers of (8.12), numerically verifying some stability boundaries in Fig. 8.16, as well as the presence of the multipliers ± 1 along some of the $\pm 45°$ lines as proved in [104, 106].

To begin with, we convert (8.12) into a first-order system according to (7.2), i.e.,

$$x'(t) = \tilde{A}(t)x(t) + \tilde{B}x(t - 2\pi) \qquad (8.13)$$

for $x(t) = (y(t), y'(t))^T \in \mathbb{R}^2$, where

$$\tilde{A}(t) = \begin{pmatrix} 0 & 1 \\ -\delta - \varepsilon \cos(t) & 0 \end{pmatrix}$$

is periodic with period $\omega = 2\pi$ (hence equal to the delay $d_1 = 2\pi$) and

$$\tilde{B} = \begin{pmatrix} 0 & 0 \\ b & 0 \end{pmatrix}$$

is constant. The relevant script myDDE_mathieu1.m contains:

```
%% MEMO LIST OF POSSIBLE PARAMETERS
%par(1)=delta
%par(2)=epsilon
%par(3)=b

%% DIMENSION OF THE DDE
d=2;

%% CURRENT TIME TERM
Atilde=@(t,d,par) [0,1;-par(1)-par(2)*cos(t),0];

%% DISCRETE DELAY TERMS
dd=2*pi;
Btilde{1}=@(t,d,par) [0,0;par(3),0];

%% DISTRIBUTED DELAY TERMS
l=[];
r=[];
Ctilde{1}=@(t,theta,d,par) [];
```

8.3 Periodic Problems in Engineering

For $\delta = 2$ and $\varepsilon = 1$, we check the presence of a secondary Hopf (or Neimark-Sacker) bifurcation [106, 127] as b (b_0 in Fig. 8.16) decreases across zero:

```
>> delta=2;epsilon=1;b=.1;
>> mu=eigTMN('myDDE_mathieu1',...
[delta,epsilon,b],0,2*pi,20,20);
>> [mu(1:2),abs(mu(1:2))]
ans =
  -0.7902 + 0.4039i   0.8874
  -0.7902 - 0.4039i   0.8874
>> delta=2;epsilon=1;b=0;
>> mu=eigTMN('myDDE_mathieu1',...
[delta,epsilon,b],0,2*pi,20,20);
>> [mu(1:2),abs(mu(1:2))]
ans =
  -0.7581 + 0.6521i   1.0000
  -0.7581 - 0.6521i   1.0000
>> delta=2;epsilon=1;b=-.1;
>> mu=eigTMN('myDDE_mathieu1',...
[delta,epsilon,b],0,2*pi,20,20);
>> [mu(1:2),abs(mu(1:2))]
ans =
  -0.7755 + 0.8756i   1.1696
  -0.7755 - 0.8756i   1.1696
```

Note the high accuracy at the Hopf point:

```
>> format long
>> delta=2;epsilon=1;b=0;
>> mu=eigTMN('myDDE_mathieu1',...
[delta,epsilon,b],0,2*pi,20,20);
>> abs(mu(1:2))
ans =
   1.000000000000001
   1.000000000000001
```

The dominant multipliers are depicted in Fig. 8.17.

Always for $\delta = 2$ and $\varepsilon = 1$, we verify the presence of the multipliers ± 1 in correspondence of the $\pm 45°$ lines in Fig. 8.16:

```
>> delta=2;epsilon=1;b=.285156917225102;
>> mu=eigTMN('myDDE_mathieu1',...
[delta,epsilon,b],0,2*pi,20,20);
>> [mu(1:4),abs(mu(1:4))]
ans =
  -1.0000              1.0000
```

Fig. 8.17 The dominant multipliers of (8.13) for $\delta = 2$ and $\varepsilon = 1$ in a secondary Hopf bifurcation for varying $b = 0.1$ (×), $b = 0$ (•) and $b = -0.1$ (o) computed by eigTMN.m with $M = N = 20$

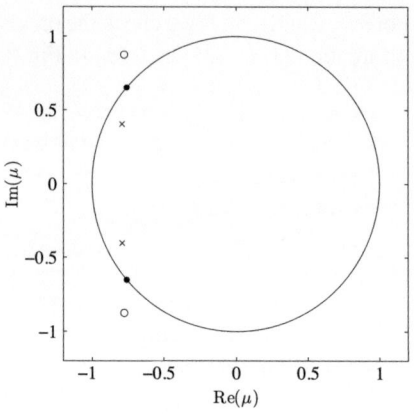

```
  -0.9450              0.9450
   0.3119 + 0.1092i    0.3304
   0.3119 - 0.1092i    0.3304
>> delta=2;epsilon=1;b=.342580621746523;
>> mu=eigTMN('myDDE_mathieu1',...
[delta,epsilon,b],0,2*pi,20,20);
>> [mu(1:4),abs(mu(1:4))]
ans =
  -1.0608              1.0608
  -1.0000              1.0000
   0.3914 + 0.0961i    0.4030
   0.3914 - 0.0961i    0.4030
>> delta=2;epsilon=1;b=.706833720464083;
>> mu=eigTMN('myDDE_mathieu1',...
[delta,epsilon,b],0,2*pi,20,20);
>> [mu(1:4),abs(mu(1:4))]
ans =
  -1.3354 + 0.4698i    1.4157
  -1.3354 - 0.4698i    1.4157
   1.0000              1.0000
   0.6731              0.6731
>> delta=2;epsilon=1;b=1.081941825617849;
>> mu=eigTMN('myDDE_mathieu1',...
[delta,epsilon,b],0,2*pi,20,20);
>> [mu(1:4),abs(mu(1:4))]
ans =
  -1.5798 + 0.8333i    1.7861
  -1.5798 - 0.8333i    1.7861
   1.4538              1.4538
   1.0000              1.0000
```

8.3 Periodic Problems in Engineering

By switching to format long, one can verify that the above values are accurate to machine precision. Indeed, the values of b for the four cases are computed through fzero.m as described in Sect. 8.2.

As a second application, we consider the following damped and delayed Mathieu equation

$$y''(t) + a_1 y'(t) + (\delta + \varepsilon \cos(\Omega t)) y(t) = b_1 y(t - d_1) + b_2 y'(t - d_2). \quad (8.14)$$

It is an instance of (4.30) in [106] with two delays. A similar equation with constant coefficients (i.e., $\varepsilon = 0$) was investigated originally in [152]. The example of stick-balancing treated in Sect. 1.4 belongs to this class of equations.

Here we aim at verifying the presence of dominant multipliers crossing the unit circle with reference to the stability chart for varying d_1 and d_2 represented in [106, Fig. 4.4 (top-right)] for $a_1 = 1.5$, $\delta = 60$, $\varepsilon = 30$, $\Omega = 2\pi$, $b_1 = -14$ and $b_2 = -1.4$, i.e., precisely equation (4.50) in [106]. A portion of the same chart is reproduced here in Fig. 8.19, obtained through eigTMN.m with $M = N = 20$ and the use of the algorithm level.m described in [41] (see also [47]). The latter implements an adaptive triangulation of the parameters plane in order to reduce the computational cost w.r.t. uniform grid contour algorithms such as MATLAB contour.m [177]. In Fig. 8.19, the line $d_2 = d_1 \pi / 2$ crosses the stability boundary at the points (A) and (B), where we verify the presence of multipliers on the unit circle.

As done before, we convert (8.14) into the first-order system according to (7.2)

$$x'(t) = \tilde{A}(t) x(t) + \tilde{B}_1 x(t - d_1) + \tilde{B}_2 x(t - d_2) \quad (8.15)$$

for $x(t) = (y(t), y'(t))^T \in \mathbb{R}^2$, where

$$\tilde{A}(t) = \begin{pmatrix} 0 & 1 \\ -\delta - \varepsilon \cos(\Omega t) & -a_1 \end{pmatrix}$$

is periodic with period $\omega = 2\pi / \Omega$ and

$$\tilde{B}_1 = \begin{pmatrix} 0 & 0 \\ b_1 & 0 \end{pmatrix}, \quad \tilde{B}_2 = \begin{pmatrix} 0 & 0 \\ 0 & b_2 \end{pmatrix}$$

are constant matrices. This is an example of analysis where the delays d_1 and d_2 are not necessarily ordered in their variation, therefore requiring the (automatic) conversion to model (7.1) as explained in Sect. 7.1. The relevant script myDDE_mathieu2.m contains:

```
%% MEMO LIST OF POSSIBLE PARAMETERS
%par(1)=a1
%par(2)=delta
%par(3)=epsilon
%par(4)=Omega
```

```
%par(5)=b1
%par(6)=b2
%par(7)=d1
%par(8)=d2

%% DIMENSION OF THE DDE
d=2;

%% CURRENT TIME TERM
Atilde=@(t,d,par) [0,1;-par(2)-par(3)*...
    cos(par(4)*t),-par(1)];

%% DISCRETE DELAY TERMS
dd=[par(7),par(8)];
Btilde{1}=@(t,d,par) [0,0;par(5),0];
Btilde{2}=@(t,d,par) [0,0;0,par(6)];

%% DISTRIBUTED DELAY TERMS
l=[];
r=[];
Ctilde{1}=@(t,theta,d,par) [];
```

The values of d_1 at the two points of bifurcation (A) and (B) in Fig. 8.18 are found again through fzero.m. The resulting multipliers are accurate to machine precision. At point (A), a secondary Hopf bifurcation occurs with the dominant complex-conjugate pair crossing the unit circle, see also Fig. 8.19 (left):

```
>> a1=1.5;delta=60;epsilon=30;
>> Omega=2*pi;b1=-14;b2=-1.4;
>> d1=.120038184259895;d2=d1*pi/2;
```

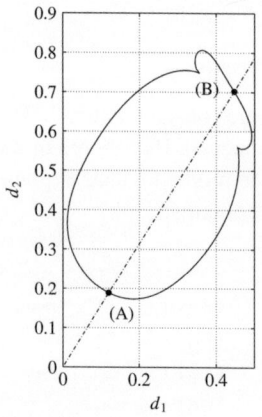

Fig. 8.18 Stability chart of (8.13) for $a_1 = 1.5$, $\delta = 60$, $\varepsilon = 30$, $\Omega = 2\pi$, $b_1 = -14$, $b_2 = -1.4$ and varying d_1 and d_2 (the *dashed-dotted line* is $d_2 = d_1\pi/2$, the stable region is outside the *closed solid curve*)

8.3 Periodic Problems in Engineering

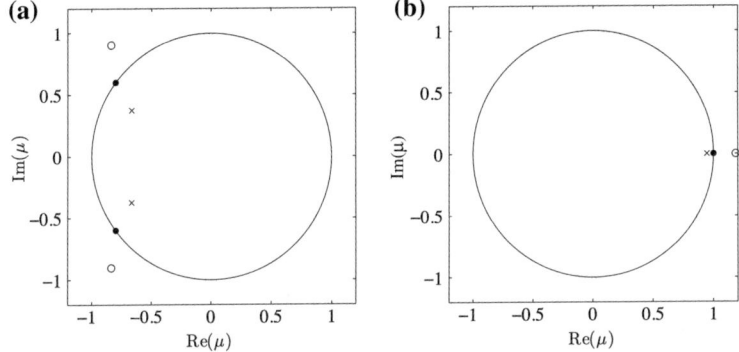

Fig. 8.19 Dominant multipliers of (8.15) for $a_1 = 1.5$, $\delta = 60$, $\varepsilon = 30$, $\Omega = 2\pi$, $b_1 = -14$, $b_2 = -1.4$ and $d_2 = d_1 \pi/2$ with $d_1 = 0.1$ (×), $d_1 = 0.120038184259895$ (•), $d_1 = 0.14$ (◦) (*left, secondary Hopf bifurcation,* (**a**) *in Fig. 8.18*) and $d_1 = 0.45$ (×), $d_1 = 0.447114615168396$ (•), $d_1 = 0.42$ (◦) (*right, cyclic-fold bifurcation,* (**b**) *in Fig. 8.18*)

```
>> mu=eigTMN('myDDE_mathieu2',...
[a1,delta,epsilon,Omega,b1,b2,d1,d2],...
0,2*pi/Omega,20,20);
>> [mu(1:2),abs(mu(1:2))]
ans =
  -0.8000 + 0.6000i   1.0000
  -0.8000 - 0.6000i   1.0000
```

At point (B), a cyclic-fold bifurcation [106, 127] occurs with the dominant multiplier crossing the unit circle at 1, see also Fig. 8.19 (right):

```
>> a1=1.5;delta=60;epsilon=30;
>> Omega=2*pi;b1=-14;b2=-1.4;
>> d1=.447114615168396;d2=d1*pi/2;
>> mu=eigTMN('myDDE_mathieu2',...
[a1,delta,epsilon,Omega,b1,b2,d1,d2],...
0,2*pi/Omega,20,20);
>> [mu(1),abs(mu(1))]
ans =
    1.0000    1.0000
```

The last Mathieu equation we consider has a distributed delay:

$$y''(t) + (\delta + \varepsilon \cos(\Omega t))y(t) = b\frac{\pi}{2}\int_{-1}^{0} \sin(\pi\theta)y(t+\theta)\,d\theta. \qquad (8.16)$$

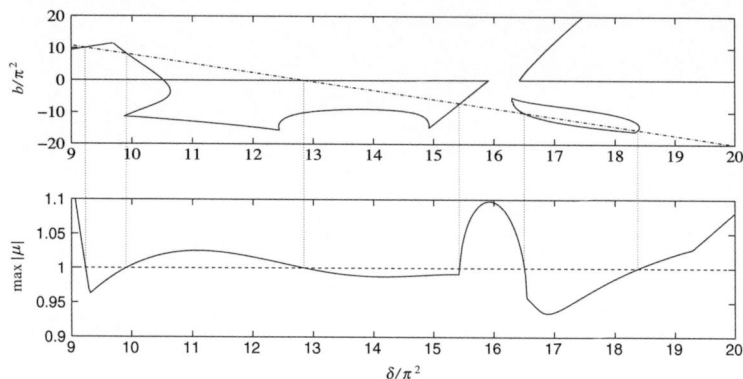

Fig. 8.20 Stability chart of (8.17) for $\varepsilon = 4\pi^2$, $\Omega = 4\pi$ and varying δ and b (*top*, the stable regions are inside the *closed solid curves*); the absolute value of the dominant multiplier of (8.17) along the *dashed-dotted line* in the *top panel* (*bottom*).

It is also investigated in [106], see (4.52) and (4.77). The autonomous case $\varepsilon = 0$ was originally analyzed in [178], while similar periodic instances appear in [47, 120].

Here we aim at monitoring the dominant multiplier as the parameters δ and b vary for fixed $\varepsilon = 4\pi^2$ and $\Omega = 4\pi$. The period is thus $\omega = 1/2$. As a reference, we reproduce in Fig. 8.20 (top) the relevant stability chart (compare with [106, Fig. 4.9 bottom-left]), where the line $b = -14/5\delta + 36$ is also sketched. In Fig. 8.20 (bottom), the behavior of the absolute value of the dominant multiplier along this line is represented. All computations are carried out through eigTMN.m with $M = N = 10$. The relevant script myDDE_mathieu3.m contains:

```
%% MEMO LIST OF POSSIBLE PARAMETERS
%par(1)=delta
%par(2)=epsilon
%par(3)=Omega
%par(4)=b

%% DIMENSION OF THE DDE
d=2;

%% CURRENT TIME TERM
Atilde=@(t,d,par) [0,1;-par(1)-par(2)*cos(par(3)*t),0];

%% DISCRETE DELAY TERMS
dd=[];
Btilde{1}=@(t,d,par) [];
```

8.3 Periodic Problems in Engineering

```
%% DISTRIBUTED DELAY TERMS
l=[1];
r=[0];
Ctilde{1}=@(t,theta,d,par) [0,0;...
    par(4)*pi/2*sin(pi*theta),0];
```

and it refers to the first-order system of DDEs rewritten according to (7.2) as

$$x'(t) = \tilde{A}(t)x(t) + \int_{-1}^{0} \tilde{C}(\theta)x(t+\theta)\,d\theta \qquad (8.17)$$

for $x(t) = (y(t), y'(t))^T \in \mathbb{R}^2$, where

$$\tilde{A}(t) = \begin{pmatrix} 0 & 1 \\ -\delta - \varepsilon\cos(\Omega t) & 0 \end{pmatrix}$$

is periodic with period $\omega = 2\pi/\Omega$ and

$$\tilde{C}(\theta) = \begin{pmatrix} 0 & 0 \\ b\dfrac{\pi}{2}\sin(\pi\theta) & 0 \end{pmatrix}.$$

References

1. Matlab documentation online: anonymous functions. http://www.mathworks.it/it/help/matlab/matlab_prog/anonymous-functions.html
2. Matlab documentation online: array vs. matrix operations. http://www.mathworks.it/it/help/matlab/matlab_prog/array-vs-matrix-operations.html
3. Matlab documentation online: cell-arrays. http://www.mathworks.it/it/help/matlab/cell-arrays.html
4. Matlab documentation online: eig. http://www.mathworks.it/it/help/matlab/ref/eig.html
5. Matlab documentation online: eigs. http://www.mathworks.it/it/help/matlab/ref/eigs.html
6. Abell, K., Elmer, C., Humphries, A.R., Van Vleck, E.S.: Computation of mixed type functional differential boundary value problems. SIAM J. Dyn. Syst. **4**(3), 755–781 (2005)
7. Abrams, P.A., Walters, C.J.: Invulnerable prey and the statics and dynamics of predator-prey interaction. Ecology **77**, 1125–1133 (1996)
8. Ambrosetti, A., Prodi, G.: A Primer of Nonlinear Analysis. Cambridge Studies in Advanced Mathematics, vol. 34. Cambridge Univeristy Press, New York (1995)
9. Anwar Sadath, K.T., Vyasarayani, C.P.: Galerkin approximations for stability of delay differential equations with time periodic coefficients. J. Comput. Nonlinear Dyn. (2014)
10. Arditi, R., Ginzburg, L.R.: Coupling in predator-prey dynamics: ratio-dependence. J. Theor. Biol. **139**, 311–326 (1989)
11. Azbelev, N., Maksimov, V.P., Rakhmatullina, L.F.: Introduction to the Theory of Functional Differential Equations: Methods and Applications. Contemporary Mathematics and Its Applications, vol. 3. Hindawi Publishing Corporation, New York (2007)
12. Azbelev, N., Simonov, P.: Stability of Differential Equations with Aftereffect. Stability and Control: Theory, Methods and Applications, vol. 20. Taylor & Francis, London (2002)
13. Bader, G.: Solving boundary value problems for functional-differential equations by collocation. Numerical Boundary Value ODEs (Vancouver, B.C., 1984), vol. 5, pp. 227–243. Birkhäuser, Boston (1985)
14. Baker, C., Ford, N.: Stability properties of a scheme for the approximate solution of a delay-integro-differential equation. Appl. Numer. Math. **9**, 357–370 (1992)
15. Batkai, A., Piazzera, S.: Semigroups for Delay Equations. Research Notes in Mathematics, vol. 10. A K Peters Ltd, Canada (2005)
16. Bayly, P.V., Halley, J.E., Mann, B.P., Davies, M.A.: Stability of interrupted cutting by temporal finite element analysis. In: Proceedings of 2001 ASME DETC. Pittsburgh (2001)
17. Beddington, J.R.: Mutual interference between parasites or predators and its effect on searching efficiency. J. Anim. Ecol. **44**, 331–340 (1975)

18. Bellen, A., Guglielmi, A., Maset, S., Zennaro, M.: Recent trends in the numerical solution of retarded functional differential equations. Acta Numer. **18**, 1–110 (2009)
19. Bellen, A., Maset, S.: Numerical solution of constant coefficient linear delay differential equations as abstract cauchy problems. Numer. Math. **84**, 351–374 (2000)
20. Bellen, A., Zennaro, M.: Numerical Methods for Delay Differential Equations. Numerical Mathemathics and Scientifing Computing Series. Oxford University Press, New York (2003)
21. Bellman, R.: Note on the derivatives with respect to a parameter of the solutions of a system of differential equations. Duke Math. **4**(10), 643–647 (1943)
22. Bellman, R.E., Cooke, K.L.: Differential-Difference Equations. Academic Press, New York (1963)
23. Beretta, E., Breda, D.: An SEIR epidemic model with constant latency time and infectious period. Math. Biosci. Eng. **8**(4), 931–952 (2011)
24. Berrut, J.P., Trefethen, L.N.: Barycentric lagrange interpolation. SIAM Rev. **46**(3), 501–517 (2004)
25. Blythe, S.P., Nisbet, R.M., Gurney, W.S.C.: Stability switches in distributed delay models. J. Math. Anal. Appl. **109**, 388–396 (1985)
26. Boese, F.G.: The stability chart for the linearized Cushing equation with a discrete delay and with gamma-distributed delays. J. Math. Anal. Appl. **140**, 510–536 (1989)
27. Bozzo, E., Breda, D., Vermiglio, R.: Characteristic roots of delay differential equations: is this the end? (2010). Poster presented at IFAC Time Delay Systems 2010, Prague, available at http://sole.dimi.uniud.it/~dimitri.breda/wp-content/uploads/2013/11/SDDE09_pos.pdf
28. Breda, D.: The infinitesimal generator approach for the computation of characteristic roots for delay differential equations using BDF methods. Technical Report RR17/2002, Department of Mathematics and Computer Science, University of Udine (2002)
29. Breda, D.: Methods for numerical computation of characteristic roots for delay differential equations: experimental comparison. Sci. Math. Jpn. **58**(2), 317–328 (2003)
30. Breda, D.: Numerical computation of characteristic roots for delay differential equations. Ph.D. thesis, in Computational Mathematics. Università di Padova (2004)
31. Breda, D.: Solution operator approximation for characteristic roots of delay differential equations. Appl. Numer. Math. **56**(3–4), 305–317 (2006)
32. Breda, D.: Nonautonomous delay differential equations in Hilbert spaces and Lyapunov exponents. Differ. Int. Equ. **23**(9–10), 935–956 (2010)
33. Breda, D.: On characteristic roots and stability charts of delay differential equations. Int. J. Robust Nonlinear **22**, 892–917 (2012)
34. Breda, D., Cusulin, C., Iannelli, M., Maset, S., Vermiglio, R.: Stability analysis of age-structured population equations by pseudospectral differencing methods. J. Math. Biol. **54**(5), 701–720 (2007)
35. Breda, D., Diekmann, O., Maset, S., Vermiglio, R.: A numerical approach for investigating the stability of equilibria for structured population models. J. Biol. Dyn. **7**(1), 4–20 (2013)
36. Breda, D., Iannelli, M., Maset, S., Vermiglio, R.: Stability analysis of the Gurtin-MacCamy model. SIAM J. Numer. Anal. **46**(2), 980–995 (2008)
37. Breda, D., Maset, S., Vermiglio, R.: Computing the characteristic roots for delay differential equations. IMA J. Numer. Anal. **24**(1), 1–19 (2004)
38. Breda, D., Maset, S., Vermiglio, R.: Pseudospectral differencing methods for characteristic roots of delay differential equations. SIAM J. Sci. Comput. **27**(2), 482–495 (2005)
39. Breda, D., Maset, S., Vermiglio, R.: Numerical computation of characteristic multipliers for linear time periodic delay differential equations. In: Manes C., Pepe P. (eds.) Time Delay Systems 2006, IFAC Proceedings Volumes, vol. 6. Elsevier (2006)
40. Breda, D., Maset, S., Vermiglio, R.: Pseudospectral approximation of eigenvalues of derivative operators with non-local boundary conditions. Appl. Numer. Math. **56**(3–4), 318–331 (2006)
41. Breda, D., Maset, S., Vermiglio, R.: An adaptive algorithm for efficient computation of level curves of surfaces. Numer. Algorithms **52**(4), 605–628 (2009a)
42. Breda, D., Maset, S., Vermiglio, R.: Numerical approximation of characteristic values of partial retarded functional differential equations. Numer. Math. **113**(2), 181–242 (2009b)

References

43. Breda, D., Maset, S., Vermiglio, R.: Computation of asymptotic stability for a class of partial differential equations with delay. J. Vib. Control **16**(7–8), 1005–1022 (2010)
44. Breda, D., Maset, S., Vermiglio, R.: Approximation of eigenvalues of evolution operators for linear retarded functional differential equations. SIAM J. Numer. Anal. **50**(3), 1456–1483 (2012a)
45. Breda, D., Maset, S., Vermiglio, R.: Computing eigenvalues of Gurtin-MacCamy models with diffusion. IMA J. Numer. Anal. **32**(3), 1030–1050 (2012b)
46. Breda, D., Maset, S., Vermiglio, R.: Numerical recipes for investigating endemic equilibria of age-structured SIR epidemics. Discret. Contin. Dyn. S **32**(8), 2675–2699 (2012c)
47. Breda, D., Maset, S., Vermiglio, R.: Pseudospectral methods for stability analysis of delayed dynamical systems. Int. J. Dyn. Control **2**, 143–153 (2014)
48. Breda, D., Maset, S., Vermiglio, R.: This book matlab codes (2014). http://sole.dimi.uniud.it/~dimitri.breda/research/software/
49. Breda, D., Van Vleck, E.S.: Approximating Lyapunov exponents and Sacker-Sell spectrum for retarded functional differential equations. Numer. Math. **126**, 225–257 (2014)
50. Breda, D., Visetti, D.: Existence, multiplicity and stability of endemic states for an age-structured S-I epidemic model. Math. Biosci. **235**, 19–31 (2012)
51. Bueler, E.: Error bounds for approximate eigenvalues of periodic-coefficient linear delay differential equations. SIAM J. Numer. Anal. **45**(6), 2510–2536 (2007)
52. Bueler, E., Averina, V., Butcher, E.: Periodic linear DDEs: collocation approximation to the monodromy operator. In: SIAM Annual Meeting. Portland (2004)
53. Butcher, E.A., Bobrenkov, O.A.: On the Chebyshev spectral continuous time approximation for constant and periodic delay differential equations. Commun. Nonlinear Sci. Numer. Simul. **16**, 1541–1554 (2011)
54. Butcher, E.A., Ma, H.T., Bueler, E., Averina, V., Szabo, Z.: Stability of linear time-periodic delay-differential equations via Chebyshev polynomials. Int. J. Numer. Meth. Eng. **59**, 895–922 (2004)
55. Butcher, J.C.: The numerical analysis of ordinary differential equations. Runge-Kutta and general linear methods. A Wiley Interscience Publication. Wiley, Chichester (1987)
56. Cantrell, R.S., Cosner, C.: On the dynamics of predator-prey models with the Beddington-DeAngelis functional response. J. Math. Anal. Appl. **257**, 206–222 (2001)
57. Canuto, C., Hussaini, M.Y., Quarteroni, A., Zang, T.: Spectral Methods. Fundamentals in single domains. Scientific Computation Series. Springer, Berlin (2006)
58. Caperon, J.: Time lag in population growth response of isochrysis galbana to a variable nitrate environment. Ecology **50**, 188–192 (1969)
59. Champneys, A., Fraser, W.B.: The "Indian rope trick" for a parametrically excited flexible rod: linearized analysis. R. Soc. Lond. Proc. Ser. A Math. Phys. **456**, 553–570 (2000)
60. Chatelin, F.: Spectral Approximation of Linear Operators. Academic Press, New York (1983)
61. Chicone, C.: Ordinary Differential Equations with Applications. Springer, New York (1999)
62. Chicone, C., Latushkin, Y.: Evolution Semigroups in Dynamical Systems and Differential Equations. No. 70 in Mathematical Surveys and Monographs. American Mathematical Society, Providence (1999)
63. Cooke, K.L., Grossman, Z.: Discrete delay, distributed delays and stability switches. J. Math. Anal. Appl. **86**, 592–627 (1982)
64. Corless, R., Gonnet, G., Hare, D., Jeffrey, D., Knuth, D.: On the Lambert W function. Adv. Comput. Math. **5**, 329–359 (1996)
65. Cosner, C., DeAngelis, D.L., Ault, J.S., Olson, D.B.: Effects of spatial grouping on the functional response of predators. Theor. Popul. Biol. **56**, 65–75 (1999)
66. Cushing, J.M.: Time delays in single species growth models. J. Math. Biol. **3**, 257–264 (1977)
67. Davis, P.J.: Interpolation & Approximation. Dover, New York (1975)
68. DeAngelis, D.L., Goldstein, R.A., Neill, R.: A model for trophic interaction. Ecology **56**, 881–892 (1975)
69. Deuflhard, P.: Newton Methods for Nonlinear Problems. Affine Invariance and Adaptive Algorithms, 2nd edn. Computational Mathematics, vol. 35. Springer, Heidelberg (2011)

70. Diekmann, O., van Gils, S.A., Verduyn Lunel, S.M., Walther, H.O.: Delay Equations—Functional, Complex and Nonlinear Analysis. Applied Mathematical Sciences, vol. 110. Springer, New York (1995)
71. Diekmann, O., Gyllenberg, M., Metz, J.A.J., Nakaoka, S., de Roos, A.M.: Daphnia revisited: local stability and bifurcation theory for physiologically structured population models explained by way of an example. J. Math. Biol. **61**(2), 277–318 (2010)
72. Driver, R.D.: Ordinary and Delay Differential Equations. Applied Mathematical Sciences, vol. 20. Springer, New York (1977)
73. Linear Algebra in Action. Graduate Studies in Mathematics, vol. 78. American Mathematical Society, Providence (2007)
74. Engel, K., Nagel, R.: One-Parameter Semigroups for Linear Evolution Equations. Graduate Texts in Mathematics, vol. 194. Springer, New York (1999)
75. Engelborghs, K., Doedel, E.: Stability of piecewise polynomial collocation for computing periodic solutions of delay differential equations. Numer. Math. **91**(4), 627–648 (2002)
76. Engelborghs, K., Luzyanina, T., In 't Hout, K.J., Roose, D.: Collocation methods for the computation of periodic solutions of delay differential equations. SIAM J. Sci. Comput. **22**(5), 1593–1609 (2000)
77. Engelborghs, K., Luzyanina, T., Roose, D.: Numerical bifurcation analysis of delay differential equations. J. Comput. Appl. Math. **125**(1–2), 265–275 (2000)
78. Engelborghs, K., Luzyanina, T., Roose, D.: Numerical bifurcation analysis of delay differential equations using DDE-BIFTOOL. ACM T. Math. Softw. **28**(1), 1–21 (2002)
79. Engelborghs, K., Luzyanina, T., Samaey, G.: DDE-BIFTOOL v. 2.00: a Matlab package for bifurcation analysis of delay differential equations. Technical Report TW330, Department of Computer Science, K. U. Leuven, Belgium (2001)
80. Engelborghs, K., Roose, D.: On stability of LMS methods and characteristic roots of delay differential equations. SIAM J. Numer. Anal. **40**(2), 629–650 (2002)
81. Erneux, T.: Applied delay differential equations. Surveys and Tutorials in the Applied Mathematical Sciences, vol. 3. Springer, New York (2009)
82. Farkas, M.: Periodic Motions. Applied Mathematical Sciences, vol. 10. Springer, New York (1994)
83. Farmer, D.: Chaotic attractors of an infinite-dimensional dynamical system. Physica D **4**, 605–617 (1982)
84. Floquet, G.: Sur les équations différentielles linéaires à coefficients périodiques. Ann. Sci. École Norm. S. **12**, 47–88 (1883)
85. Franceschetti, A., Pugliese, A., Breda, D.: Multiple endemic states in age-structured SIR epidemic models. Math. Biosci. Eng. **9**(3), 577–599 (2012)
86. Golub, G., Van Loan, C.: Matrix Computations. Johns Hopkins Studies in Mathematical Sciences, 4th edn. Johns Hopkins University Press, Baltimore (2013)
87. Gottlieb, D., Orszag, S.: Numerical Analysis of Spectral Methods: Theory and Applications. CBMS-NSF Regional Conference Series in Applied Mathematics, vol. 26. SIAM, Philadelphia (1977)
88. Gronwall, T.H.: Note on the derivatives with respect to a parameter of the solutions of a system of differential equations. Ann. Math. **2**(20), 292–296 (1919)
89. Hadamard, J.: Sur les problémes aux dèrivés partielles et leur signification physique. Princeton Univ. Bull. **13**, 49–52 (1902)
90. Hairer, E., Nörsett, S.P., Wanner, G.: Solving Ordinary Differential Equations I: Nonstiff Problems, 2nd edn. Computational Mathematics, vol. 8. Springer, Berlin (1993)
91. Hale, J.K.: Theory of Functional Differential Equations, 1st edn. Applied Mathematical Sciences, vol. 99. Springer, New York (1977)
92. Hale, J.K., Koçak, H.: Dynamics and Bifurcations. Tam Series, vol. 3. Springer, New York (1991)
93. Hale, J.K., Verduyn Lunel, S.M.: Introduction to Functional Differential Equations, 2nd edn. Applied Mathematical Sciences, vol. 99. Springer, New York (1993)

References

94. Hangelbroek, R.J., Kaper, H.G., Leaf, G.K.: Collocation methods for integro-differential equations. SIAM J. Numer. Anal. **14**(3), 377–390 (1977)
95. Hayes, N.D.: Roots of the transcendental equation associated with a certain difference-differential equation. J. Lond. Math. Soc. **25**, 226–231 (1950)
96. Henderson, J.: Boundary Value Problems for Functional-differential Equations. World Scientific Publishing Co., River Edge (1995)
97. Hsu, C.S., Bhatt, S.J.: Stability charts for second-order dynamical systems with time lag. J. Appl. Mech. **33**(1), 119–124 (1966)
98. Hutchinson, G.E.: Circular causal systems in ecology. Ann. N.Y. Acad. Sci. **50**, 221–246 (1948)
99. Hwang, Z.W.: Global analysis of the predator-prey system with Beddington-DeAngelis functional response. J. Math. Anal. Appl. **281**, 395–401 (2003)
100. Hwang, Z.W.: Uniqueness of limit cycles of the predator-prey system with Beddington-DeAngelis functional response. J. Math. Anal. Appl. **290**, 113–122 (2004)
101. Ince, E.L.: Ordinary Differential Equations. Longmans, Green and Co., London (1926)
102. Insperger, T.: Stick balancing with reflex delay in case of parametric forcing. Commun. Nonlinear Sci. **16**, 2160–2168 (2011)
103. Insperger, T., Stépán, G.: Semi-discretization method for delayed systems. Int. J. Numer. Meth. Eng. **55**, 503–518 (2002)
104. Insperger, T., Stépán, G.: Stability chart for the delayed Mathieu equation. R. Soc. Lond. Proc. Ser. A Math. Phys. Eng. Sci. **458**(2024), 1989–1998 (2002)
105. Insperger, T., Stépán, G.: Updated semi-discretization method for periodic delay-differential equations with discrete delay. Int. J. Numer. Meth. Eng. **61**, 117–141 (2004)
106. Insperger, T., Stépán, G.: Semi-discretization for Time-delay Systems—Stability and Engineering Applications. Applied Mathematical Sciences, vol. 178. Springer, New York (2011)
107. Insperger, T., Stépán, G., Turi, J.: On the higher-order semi-discretizations for periodic delayed systems. J. Sound Vib. **313**, 334–341 (2008)
108. Ipsen, I., Rehman, R.: Perturbation bounds for determinants and characteristic polynomials. SIAM J. Matrix Anal. Appl. **30**(2), 762–776 (2008)
109. Jarlebring, E.: The Spectrum of delay-differential equations: numerical methods, stability and perturbation. Ph.D. Thesis, Institute Computational Mathematics, TU Braunschweig (2008)
110. Jarlebring, E.: Critical delays and polynomial eigenvalue problems. J. Comput. Appl. Math. **224**(1), 296–306 (2009)
111. Jarlebring, E., Damm, T.: The Lambert W function and the spectrum of some multidimensional time-delay systems. Automatica **43**(12), 2124–2128 (2007)
112. Jarlebring, E., Hochstenbach, M.E.: Polynomial two-parameter eigenvalue problems and matrix pencil methods for stability of delay-differential equations. Linear Algebra Appl. **431**(3–4), 369–380 (2009)
113. Jarlebring, E., Meerbergen, K., Michiels, W.: An Arnoldi method with structured starting vectors for the delay eigenvalue problem. In: Vyhlidal T., Zitek P. (eds.) Time Delay Systems 2010, IFAC Proceedings Volumes, vol. 9. Elsevier, Amsterdam (2010)
114. Jarlebring, E., Meerbergen, K., Michiels, W.: A Krylov method for the delay eigenvalue problem. SIAM J. Sci. Comput. **32**(6), 3278–3300 (2010)
115. Jarlebring, E., Michiels, W., Meerbergen, K.: The infinite Arnoldi method and an application to time-delay systems with distributed delays. In: Sipahi, R., Vyhlidal, T., Niculescu, S.I., Pepe, P. (eds.) Time Delay Systems—Methods, Applications and New Trends. LNCIS, vol. 423. Springer, New York (2012)
116. Jarlebring, E., Vanbiervliet, J., Michiels, W.: Characterizing and computing the $\mathcal{H}_2$ norm of time-delay systems by solving the delay Lyapunov equation. IEEE Trans. Autom. Control **56**(4), 814–825 (2011)
117. Kalmár-Nagy, T.: Stability analysis of delay-differential equations by the method of steps and inverse Laplace transforms. Differ. Equ. Dyn. Syst. **17**(1–2), 185–200 (1951)
118. Kapitsa, P.L.: Dynamic stability of a pendulum with an oscillating point of suspension. Zh. Eksper. Teoret. Fiz. **21**, 588–597 (1951)

119. Khasawneh, F.A., Mann, B.P.: A spectral element approach for the stability of delay systems. Int. J. Numer. Meth. Eng. **87**(6), 566–592 (2011)
120. Khasawneh, F.A., Mann, B.P.: Stability of delay integro-differential equations using a spectral element method. Math. Comput. Model. **54**, 2493–2503 (2011)
121. Kolmanovskii, V.B., Myshkis, A.: Applied Theory of Functional Differential Equations. Mathematics and its Applications (Soviet Series), vol. 85. Kluver Academic Press, The Netherlands (1992)
122. Kolmanovskii, V.B., Nosov, V.R.: Stability of Functional Differential Equations. Mathematics in Science and Engineering, vol. 180. Academic Press, London (1986)
123. Krasovskii, N.: Stability of Motion. Moscow (1959) [English translation]. Stanford University Press, California (1963)
124. Kress, R.: Linear Integral Equations. Applied Mathematical Sciences, vol. 82. Springer, New York (1989)
125. Kuang, J., Cong, Y.: Stability of Numerical Methods for Delay Differential Equations. Science Press, Beijing (2005)
126. Kuang, Y.: Delay Differential Equations with Application in Population Dynamics. Dynamics in Science and Engineering, vol. 191. Academic Press, New York (1993)
127. Kuznetsov, Y.A.: Elements of Applied Bifurcation Theory, 2nd edn. Applied Mathematical Sciences, vol. 112. Springer, New York (1998)
128. Lambert, J.D.: Numerical Methods For Ordinary Differential Equations. Wiley, Britain (1991)
129. Langer, R.E.: The asymptotic location of the roots of a certain transcendental equation. T. Am. Math. Soc. **31**, 837–844 (1929)
130. Levi, M.: Stability of the inverted pendulum: a topological explanation. SIAM Rev. **30**, 639–644 (1988)
131. Liu, L., Kalmár-Nagy, T.: High dimensional harmonic balance analysis of second-order delay-differential equations. J. Vib. Control **16**(7–8), 1189–1208 (2010)
132. Liu, S., Beretta, E.: Stage-structured predator-prey model with the Beddington-DeAngelis functional response. SIAM J. Appl. Math. **66**, 1101–1129 (2006)
133. Liu, S., Beretta, E., Breda, D.: Predator-prey model of Beddington-DeAngelis type with maturation and gestation delays. Nonlinear Anal-Real **11**, 4072–4091 (2010)
134. Liu, Z., Yuan, R.: Stability and bifurcation in a delayed predator-prey system with Beddington-DeAngelis functional response and stage structure. J. Math. Anal. Appl. **296**, 521–537 (2004)
135. Lotka, A.J.: Elements of Physical Biology. Williams and Wilkins Company, New York (1925)
136. Luzyanina, T., Engelborghs, K., Lust, K., Roose, D.: Computation, continuation and bifurcation analysis of periodic solutions of delay differential equations. Intern. J. Bifur. Chaos Appl. Sci. Eng. **7**(11), 2547–2560 (1997)
137. Ma, H., Butcher, E.A.: Stability of elastic columns with periodic retarded follower forces. J. Sound Vib. **286**, 849–867 (2005)
138. Mackey, M.C., Glass, L.: Oscillations and chaos in physiological control systems. Science **197**, 287–289 (1977)
139. Malakhovski, E., Mirkin, L.: On stability of second-order quasi-polynomials with a single delay. Automatica **42**, 1041–1047 (2006)
140. Mallet-Paret, J.: The Fredholm alternative for functional-differential equations of mixed type. J. Dyn. Differ. Equ. **11**(1), 1–47 (1999)
141. Mallet-Paret, J., Verduyn Lunel, S.: Mixed-type functional differential equations, holomorphic factorization and applications. In: EQUADIFF 2003, pp. 73–89. World Scientific Publications, Singapore (2005)
142. Malthus, T.R.: An Essay on the Principle of Population. J. Johnson, London (1798)
143. Mann, B.P., Patel, B.R.: Stability of delay equations written as state space models. J. Vib. Control **16**, 1067–1085 (2010)
144. Mathieu, E.: Mémoire sur le mouvement vibratoire d'une membrane de forme elliptique. J. Math. Pure Appl. **13**, 137–203 (1868)
145. Meyer, C.: Matrix Analysis and Applied Linear Algebra. SIAM, New York (2000)

References

146. Michiels, W., Jarlebring, E., Meerbergen, K.: Krylov-based model order reduction of time-delay systems. SIAM J. Matrix Anal. Appl. **32**(4), 1399–1421 (2011)
147. Michiels, W., Niculescu, S.I.: Stability and Stabilization of Time-delay Systems. An Eigenvalue Based Approach. Advances in Design and Control, vol. 12. SIAM, Philadelphia (2007)
148. Morărescu, C.I., Niculescu, S.I., Gu, K.Q.: Stability crossing curves of shifted gamma-distributed delay systems. SIAM J. Appl. Dyn. Syst. **6**, 475–493 (2007)
149. Narduzzi, C.: Determinazione efficiente della stabilità di equilibri di equazioni differenziali con ritardo. Master's thesis, University of Udine (in italian) (2013)
150. Neimark, J.I.: D-subdivision and spaces of quasi-polynomials. Prikl. Mat. Mekh. **13**, 349–380 (1949)
151. Niculescu, S.I.: Delay Effects on Stability: A Robust Control Approach. No. 269 in TLNCIS. Monograph Springer, London (2001)
152. Olgac, N., Ergenc, A.F., Sipahi, R.: "Delay scheduling": a new concept for stabilization in multiple delay systems. J. Vib. Control **11**, 1159–1172 (2005)
153. Olgac, N., Sipahi, R.: An exact method for the stability analysis of time delayed LTI systems. IEEE Trans. Autom. Control **47**(5), 793–797 (2002)
154. Olgac, N., Sipahi, R.: The cluster treatment of characteristic roots and the neutral type time-delayed systems. J. Dyn. Syst.-Trans. ASME **127**, 88–97 (2005)
155. Paussa, A., Conzatti, F., Breda, D., Vermiglio, R., Esseni, D., Palestri, P.: Pseudo-spectral methods for the modelling of quantization effects in nanoscale MOS transistors. IEEE Trans. Electron Dev. (2010) In press
156. Pazy, A.: Semigroups of Linear Operators and Applications to Partial Differential Equations. Applied Mathematical Sciences, vol. 44. Springer, New York (1983)
157. Pearl, R., Reed, L.J.: On the rate of growth of the population of the United States since 1790 and its mathematical representation. Proc. Natl. Acad. Sci. **6**, 275–288 (1920)
158. van der Pol, F., Strutt, M.J.O.: On the stability of the solutions of Mathieu's equation. Philos. Mag. J. Sci. **5**, 18–38 (1928)
159. Pólya, G.: Geometrisches über die verteilung der nullstellen gewisser ganzer transzendenter funktionen. Sitzungsber. Bayer. Akad. **50**, 285–290 (1920)
160. Pontryagin, L.S.: On the zeros of some elementary transcendental functions. Izv. Akad. Nauk. SSSR **6**, 115–134 (1942)
161. Priestley, H.A.: Introduction to Complex Analysis. Oxford University Press, New York (1990)
162. Richard, J.P.: Time-delay systems: an overview of some recent advances and open problems. Automatica **39**, 1667–1694 (2003)
163. Rivlin, T.: An Introduction to the Approximation of Functions. Blaisdell, Waltham (1969)
164. Röst, G.: Neimark-Sacker bifurcation for periodic delay differential equations. Nonlinear Anal. T.M.A. **60**(5), 1025–1044 (2005)
165. Rudin, W.: Real and Complex Analysis. McGraw-Hill Book Co., New York (1987)
166. Rustichini, A.: Functional differential equations of mixed type: The linear autonomous case. J. Dyn. Differ. Equ. **1**(2), 121–143 (1989a)
167. Rustichini, A.: Hopf bifurcation for functional-differential equations of mixed type. J. Dyn. Differ. Equ. **1**(2), 145–177 (1989b)
168. Ryan, S., Knechtel, C., Getz, W.: Ecological cues, gestation length and birth timing in african buffalo (Syncerus caffer). Behav. Ecol. **18**, 635–644 (2007)
169. Schwengeler, E.: Geometrisches ber die verteilung der nullstellen spezieller ganzer funktionen (exponentialsummen). Doctoral and Habilitation Theses. Zurich (1925). doi:10.3929/ethz-a-000092005
170. Sieber, J., Szalai, R.: Characteristic matrices for linear periodic delay differential equations. SIAM J. Appl. Dyn. Syst. **10**, 129–147 (2011)
171. Silva, G.J., Datta, A., Bhattacharyya, S.P.: New results on the syntehsis of PID controllers. IEEE Trans. Autom. Control **47**, 241–252 (2002)
172. Sipahi, R., Fazelinia, K., Olgac, N.: Generalization of cluster treatment of characteristic roots for robust stability of multiple time-delayed systems. Int. J. Robust Nonlinear **18**(14), 1430–1449 (2008)

173. Sipahi, R., Olgac, N., Breda, D.: Complete stability map of neutral type first order—two time delay systems. In: American Control Conference, pp. 4933–4938. IEEE Press (2007)
174. Sipahi, R., Olgac, N., Breda, D.: A stability study on first order neutral systems with three rationally independent time delays. Int. J. Syst. Sci. **41**(12), 1445–1455 (2010)
175. Skalski, G.T., Gilliam, J.F.: Functional responses with predator interference: viable alternatives to the Holling type II model. Ecology **82**, 3083–3092 (2001)
176. Smith, H.L.: An Introduction to Delay Differential Equations with Applications to the Life Sciences. Texts in Applied Mathematics, vol. 57. Springer, New York (2011)
177. Snyder, W.V.: Algorithm 531: contour plotting [J6]. ACM Trans. Math. Softw. **4**(3), 290–294 (1978)
178. Stépán, G.: Retarded Dynamical Systems. Longman, Harlow (1989)
179. Stephenson, A.: On a new type of dynamical stability. Mem. Proc. Manch. Lit. Philos. Soc. **52**, 1–10 (1908)
180. Stuart, A.M., Humphries, A.R.: Dynamical systems and numerical analysis. Cambridge Monographs on Applied and Computational Mathematics. Cambridge University Press, Cambridge (1996)
181. Szalai, R., Stépán, G., Hogan, S.J.: Continuation of bifurcations in periodic delay-differential equations using characteristic matrices. SIAM J. Sci. Comput. **28**(4), 1301–1317 (2006)
182. Tamarkin, J.D.: Some general problems of the theory of ordinary linear differential equations and expansion of an arbitrary function in series of fundamental functions. Math. Z. **1**, 1–54 (1928)
183. Teschl, G.: Ordinary Differential Equations and Dynamical Systems. Graduate Studies in Mathematics, vol. 140. American Mathematical Society, Providence (2012)
184. Trefethen, L.N.: Spectral Methods in MATLAB. Software—Environment—Tools Series. SIAM, Philadelphia (2000)
185. Trefethen, L.N.: Is Gauss quadrature better than Clenshaw-Curtis? SIAM Rev. **50**(1), 67–87 (2008)
186. Trefethen, L.N.: Approximation Theory and Approximation Practice. Other Titles in Applied Mathematics, vol. 128. SIAM, Philadelphia (2013)
187. Trevisan, F., Specogna, R., Esseni, D., Paussa, A., Breda, D., Vermiglio, R.: Comparison between pseudospectral and discrete geometric methods for modelling quantization effects in nanoscale electron devices. IEEE Trans. Magn. **48**(2), 203–206 (2012)
188. Van Loan, C.: The ubiquitous kronecker product. J. Comput. Appl. Math. **123**(1–2), 85–100 (2000)
189. Vanbiervliet, J., Michiels, W., Jarlebring, E.: Using spectral discretization for the optimal $\mathcal{H}_2$ design of time-delay systems. Int. J. Control **84**(2), 228–241 (2011)
190. Verheyden, K., Lust, K.: A Newton-Picard collocation method for periodic solutions of delay differential equations. BIT **45**(3), 605–625 (2005)
191. Verheyden, K., Luzyanina, T., Roose, D.: Efficient computation of characteristic roots of delay differential equations using LMS methods. J. Comput. Appl. Math. **214**(1), 209–226 (2008)
192. Verhulst, P.F.: Notice sur la loi que la population poursuit dans son accroissement. Correspondence Mathématique et Physique. A, vol. 10. Quetelet, Bruxelles (1838)
193. Verhulst, P.F.: Recherches mathématiques sur la loi d'accroissement de la population. Nouv. Acad. R. Sci. Lett. B.-Arts Belg. **18**, 1–45 (1845)
194. Vyhlídal, T.: Analysis and Synthesis of Time Delay System Spectrum. Ph.D. Thesis, PhD in Control and Science Engineering, Czech Technical University in Prague, Czech (2003)
195. Vyhlídal, T., Zítek, P.: Mapping the spectrum of a retarded time-delay system utilizing root distribution features. In: Manes, C., Pepe, P. (eds.) Time Delay Systems, vol. 6. Elsevier, Amsterdam (2006)
196. Vyhlídal, T., Zítek, P.: Mapping based algorithm for large-scale computation of quasi-polynomial zeros. IEEE Trans. Autom. Control **54**(1), 171–177 (2009)
197. Wahi, P.: A Study of Delay Differential Equations with Applications to Machine Tool Vibrations. Ph.D. Thesis, Indian Institute of Science, Bangalore (2005)

References

198. Wahi, P., Chatterjee, A.: Galerkin projections for delay differential equations. J. Dyn. Syst.-Trans. ASME **127**, 80–87 (2005)
199. Wang, L., Xu, R., Feng, G.: Stability and Hopf bifurcation of a predator-prey system with time delay and Holling type-II functional response. Int. J. Biomath. **2**, 139–149 (2009)
200. Wilder, C.E.: Expansion problems of ordinary linear differential equations with auxiliary conditions at more than two points. Trans. Am. Math. Soc. **18**, 415–442 (1917)
201. Wilkinson, J.H.: The Algebraic Eigenvalue Problem. Clarendon Press, Oxford (1965)
202. Wilkinson, J.H.: The perfidious polynomial. In: Golub, G.H. (ed.) Studies in Numerical Analysis, Studies in Mathematics, vol. 24, pp. 1–28. Mathematical Association of America, Washington, D.C. (1984)
203. Wright, E.M.: The linear difference-differential equation with constant coefficients. P. Roy. Soc. Edinb. A **62**, 387–393 (1949)
204. Wu, J.: Theory and Applications of Partial Functional Differential Equations. Applied Mathematical Sciences, vol. 119. Springer, New York (1996)
205. Wu, Z., Michiels, W.: Reliably computing all characteristic roots of delay differential equations in a given right half plane using a spectral method. J. Comput. Appl. Math. **236**, 2499–2514 (2012)
206. Yi, S.: Time-Delay Systems: Analysis and Control Using the Lambert W Function. Ph.D. Thesis, PhD in Mechanical Engineering, University of Michigan, Michigan (2009)
207. Yi, S., Nelson, P.W., Ulsoy, A.G.: Time-delay Systems: Analysis and Control Using the Lambert W Function. World Scientific, Singapore (2010)

Series Editors' Biographies

Tamer Başar is with the University of Illinois at Urbana-Champaign, where he holds the academic positions of Swanlund Endowed Chair, Center for Advanced Study Professor of Electrical and Computer Engineering, Research Professor at the Coordinated Science Laboratory, and Research Professor at the Information Trust Institute. He received the B.S.E.E. degree from Robert College, Istanbul, and the M.S., M.Phil, and Ph.D. degrees from Yale University. He has published extensively in systems, control, communications, and dynamic games, and has current research interests that address fundamental issues in these areas along with applications such as formation in adversarial environments, network security, resilience in cyber-physical systems, and pricing in networks.

In addition to his editorial involvement with these *Briefs*, Başar is also the Editor-in-Chief of *Automatica*, Editor of two Birkhäuser Series on *Systems & Control* and *Static & Dynamic Game Theory*, the Managing Editor of the *Annals of the International Society of Dynamic Games* (ISDG), and member of editorial and advisory boards of several international journals in control, wireless networks, and applied mathematics. He has received several awards and recognitions over the years, among which are the Medal of Science of Turkey (1993); Bode Lecture Prize (2004) of IEEE CSS; Quazza Medal (2005) of IFAC; Bellman Control Heritage Award (2006) of AACC; and Isaacs Award (2010) of ISDG. He is a member of the US National Academy of Engineering, Fellow of IEEE and IFAC, Council Member of IFAC (2011-14), a past president of CSS, the founding president of ISDG, and president of AACC (2010–11).

Antonio Bicchi is Professor of Automatic Control and Robotics at the University of Pisa. He graduated at the University of Bologna in 1988 and was a postdoc scholar at M.I.T. A.I. Lab between 1988 and 1990.

His main research interests are in:

- dynamics, kinematics and control of complex mechanichal systems, including robots, autonomous vehicles, and automotive systems;
- haptics and dextrous manipulation; and

- theory and control of nonlinear systems, in particular hybrid (logic/dynamic, symbol/signal) systems.

He has published more than 300 papers on international journals, books, and refereed conferences.

Professor Bicchi currently serves as the Director of the Interdepartmental Research Center "E. Piaggio" of the University of Pisa, and President of the Italian Association or Researchers in Automatic Control. He has served as Editor in Chief of the Conference Editorial Board for the IEEE Robotics and Automation Society (RAS), and as Vice President of IEEE RAS, Distinguished Lecturer, and Editor for several scientific journals including the *International Journal of Robotics Research*, the *IEEE Transactions on Robotics and Automation*, and *IEEE RAS Magazine*. He has organized and co-chaired the first WorldHaptics Conference (2005), and Hybrid Systems: Computation and Control (2007). He is the recipient of several best paper awards at various conferences, and of an Advanced Grant from the European Research Council. Antonio Bicchi has been an IEEE Fellow since 2005.

Miroslav Krstic holds the Daniel L. Alspach chair and is the founding director of the Cymer Center for Control Systems and Dynamics at University of California, San Diego. He is a recipient of the PECASE, NSF Career, and ONR Young Investigator Awards, as well as the Axelby and Schuck Paper Prizes. Professor Krstic was the first recipient of the UCSD Research Award in the area of engineering and has held the Russell Severance Springer Distinguished Visiting Professorship at UC Berkeley and the Harold W. Sorenson Distinguished Professorship at UCSD. He is a Fellow of IEEE and IFAC. Professor Krstic serves as Senior Editor for *Automatica* and *IEEE Transactions on Automatic Control* and as Editor for the Springer series *Communications and Control Engineering*. He has served as Vice President for Technical Activities of the IEEE Control Systems Society. Krstic has co-authored eight books on adaptive, nonlinear, and stochastic control, extremum seeking, control of PDE systems including turbulent flows, and control of delay systems.

MIX
Papier aus verantwortungsvollen Quellen
Paper from responsible sources
FSC® C105338

If you have any concerns about our products,
you can contact us on
ProductSafety@springernature.com

In case Publisher is established outside the EU,
the EU authorized representative is:
**Springer Nature Customer Service Center GmbH
Europaplatz 3, 69115 Heidelberg, Germany**

Printed by Libri Plureos GmbH
in Hamburg, Germany

MIX
Papier aus verantwortungsvollen Quellen
Paper from responsible sources
FSC® C105338

If you have any concerns about our products,
you can contact us on
ProductSafety@springernature.com

In case Publisher is established outside the EU,
the EU authorized representative is:
**Springer Nature Customer Service Center GmbH
Europaplatz 3, 69115 Heidelberg, Germany**

Printed by Libri Plureos GmbH
in Hamburg, Germany